SpringerBriefs in Education

We are delighted to announce SpringerBriefs in Education, an innovative product type that combines elements of both journals and books. Briefs present concise summaries of cutting-edge research and practical applications in education. Featuring compact volumes of 50 to 125 pages, the SpringerBriefs in Education allow authors to present their ideas and readers to absorb them with a minimal time investment. Briefs are published as part of Springer's eBook Collection. In addition, Briefs are available for individual print and electronic purchase.

SpringerBriefs in Education cover a broad range of educational fields such as: Science Education, Higher Education, Educational Psychology, Assessment & Evaluation, Language Education, Mathematics Education, Educational Technology, Medical Education and Educational Policy.

SpringerBriefs typically offer an outlet for:

- An introduction to a (sub)field in education summarizing and giving an overview of theories, issues, core concepts and/or key literature in a particular field
- A timely report of state-of-the art analytical techniques and instruments in the field of educational research
- A presentation of core educational concepts
- An overview of a testing and evaluation method
- A snapshot of a hot or emerging topic or policy change
- An in-depth case study
- A literature review
- A report/review study of a survey
- An elaborated thesis

Both solicited and unsolicited manuscripts are considered for publication in the SpringerBriefs in Education series. Potential authors are warmly invited to complete and submit the Briefs Author Proposal form. All projects will be submitted to editorial review by editorial advisors.

SpringerBriefs are characterized by expedited production schedules with the aim for publication 8 to 12 weeks after acceptance and fast, global electronic dissemination through our online platform SpringerLink. The standard concise author contracts guarantee that:

- an individual ISBN is assigned to each manuscript
- each manuscript is copyrighted in the name of the author
- the author retains the right to post the pre-publication version on his/her website or that of his/her institution

Julie M. Davis · Sue Elliott · Eva Ärlemalm-Hagsér

Early Childhood Education for Sustainability

A Short History

Springer

Julie M. Davis
Queensland University of Technology
Brisbane, QLD, Australia

Eva Ärlemalm-Hagsér
Division of Early Childhood Education
School of Education, Culture
and Communication (UKK)
Mälardalen University
Västerås, Sweden

Sue Elliott
University of New England
Armidale, NSW, Australia

University of Cincinnati
Cincinnati, OH, USA

ISSN 2211-1921 ISSN 2211-193X (electronic)
SpringerBriefs in Education
ISBN 978-981-97-8404-2 ISBN 978-981-97-8405-9 (eBook)
https://doi.org/10.1007/978-981-97-8405-9

This Springer imprint is published by the registered company Springer Nature Singapore Pte Ltd.
The registered company address is: 152 Beach Road, #21-01/04 Gateway East, Singapore 189721, Singapore

Preface: Positioning Ourselves and the Text: Journeys and Drivers

'Playing for Keeps'

Julie Davis

I began to see myself as an early childhood environmental/sustainability educator in the mid-1990s. At the time, I was working as an early childhood teacher educator in the School of Early Childhood, Queensland University of Technology (QUT), Australia. Prior to this, I had experiences living in Papua New Guinea (PNG), and some travel in Africa. After returning to Australia, I took a primary teacher position in an Aboriginal community, then later worked in teacher education at Batchelor in the Northern Territory (NT)—a teacher education institution for Aboriginal people wishing to work in their own, often very isolated communities in the NT. These experiences led me to undertake a Bachelor of Environmental Studies, a multidisciplinary undergraduate degree, with an emphasis on studies across the natural sciences and the social sciences. What these formative experiences did was reshape my values—from privileged, white settler perspectives on Australian and global histories and cultures—towards more critical standpoints. These included social and cultural norms, mainstream educational practices, views about environment and conventional politics that leave a lot to be desired in creating change for sustainability.

At the end of this degree, I wasn't sure how or where I could use my new learning and skills. At the time I was teaching part-time at QUT (I now had two young children) with a cohort of Aboriginal and Torres Straits Islander student teachers, then full time with early childhood science and social science student teachers. These work experiences combined with a Masters in Environmental Education, gave me the 'ahh hah' moment where I could bring my love of science, (especially the natural sciences), the social sciences, and intercultural learning and teaching, into the emerging sub-field of *Early Childhood Environmental Education* (ECEE). The pedagogies of Environmental Education (EE) and Early Childhood Education (ECE) seemed such a good fit that I specialised in teaching, writing, and research in ECEE. As there were few people working in this area, I followed Sue Elliott's lead in Victoria and, with Noeleen Rowntree, school principal in a Queensland environmental centre, we established the then-Queensland Early Childhood Environmental Education Network (QECEEN) in 1996. The rest is history!

It was in the early 1990s, both as teacher educators, that Sue and I started to liaise on strategies for growing the field and began to present at Australian conferences on the topic of ECEE. Later, we extended our networks internationally. Thus, in 2007, I personally met Eva Ärlemalm-Hagsér from Sweden though we had been in email contact. I was also fortunate—through my university teaching and service, to work in educational settings such as India, Singapore, Papua New Guinea, and China that have continued to sharpen my understandings of sustainability and EfS/ECEfS in globally diverse contexts. As our local, national, and international networks strengthened and expanded, it became obvious that if we wanted people to understand a role for ECEfS, then we were the obvious ones to lead such efforts, having the encouragement and resources of universities to back our ideas and initiatives, in addition to our own stubbornness. This has led to a wide range of projects over the years, including for example, the textbook *Young Children and the Environment: Early Education for Sustainability,* now in its third edition with Sue Elliott as co-author, based on successful teaching and outreach strategies pursued over the decades.

However, while we are the co-authors of this little history book, we do not see ourselves as working in isolation. The field has been and is being built on the efforts of thousands of children, teachers, teacher educators, policymakers, families, and communities, many of whom we have never been directly connected with, but who are contributing to this ECEfS global movement. We are just lucky enough to have been given the possibility by our publisher to write this story from own perspectives.

Sue Elliott

I initially trained as a preschool teacher in the 1970s in Melbourne, Australia. Over the decades, I have worked in the early childhood sector in various guises as an early childhood teacher, consultant, author, lecturer, and researcher. After a somewhat eventful first year of preschool teaching, I stepped sideways into science, a long-term interest. I completed bachelor's and master's degrees in biological sciences, specifically microbiology. A combination of early childhood education and science studies was somewhat unique at the time and offered opportunities to work with zoos, aquaria, botanic gardens, and museums to shift some perceptions about relevant programs and pedagogies for young children. Much of my tertiary teaching was in early childhood science education, and increasingly, I was able to bring a sustainability lens to this teaching. Much later, I completed a Doctorate of Education at the University of New England, researching the intersection between outdoor playspaces and education for sustainability.

On reflection, a pivotal early event was my first *Australian Association for Environmental Education* (AAEE) conference in 1986, a lecture on deep ecology was a light bulb moment. This prompted a shift in my anthropocentric lens and attempts to reconcile my earlier reductionist science studies with global interdependencies and sustainable futures. I began wondering about the potential for early childhood education within environmental education, as only school-based initiatives existed at the time. Along with some like-minded colleagues, we established our own network *Environmental Education in Early Childhood* (EEEC Vic Inc) in 1992 to support early childhood educators seeking to implement environmental education. Network gatherings at early childhood centres exploring sustainable practices were a feature and we created sustainable play materials such as wooden sorting trays, rope dolls, and gravel bags as alternatives, when catalogues of manufactured play materials predominated. I then connected with the Victorian Association for Environmental Education and began writing articles in their professional journal *Eingana*, the first articles were written with some trepidation! There were also opportunities to conduct professional learning sessions as interest in Environmental Education grew, yet misperceptions persisted that environmental issues were well beyond the grasp of young children. My research and publication interests have extended over the years to outdoor playspaces and nature play programs, but always promoting sustainable worldviews have been foundational. Looking back, I have been writing about and advocating for ECEfS since the 1990s, this was recognised by the Australian Association for Environmental Education Fellowship Award in 2023. There is still much to do in this arena which I now leave to others.

EEEC Vic Inc also inspired other state-based networks, and it was particularly through the Queensland Early Childhood Sustainability network established in 1995 that I met Julie Davis, so we have enjoyed a lengthy period of collaborative work and friendship. As Julie moved onto the international stage by establishing the Transnational Dialogues in ECEfS Research (TND) in 2010, we connected with Scandinavian colleagues including Eva Ärlemalm-Hagsér. We have now co-authored publications and co-presented at national and international conferences many times. Along the way, we have consciously recruited others with similar interests in progressing ECEfS. Our three-way collaborations and mutual support over the last decade have cemented our commitment to young children's sustainable futures (Elliott, Ärlemalm-Hagsér & Davis, 2020). We are no longer fringe dwellers in this commitment, and as time passes, we see it as critical to reflect on and document the achievements of many in this globally important endeavour.

Eva Ärlemalm-Hagsér

In 1982, when I first qualified as a preschool teacher—the first of five children to study at university—content related to nature, environmental education, and outdoor play and learning was already important in Swedish preschool education. As a teacher working with children aged 1–6 years, this orientation was common everyday preschool practice. In 1992, when Sweden ratified Agenda 21, the municipality where I worked initiated sustainable development certification. This was organised on three levels—bronze, silver, and gold. Over the period 1994–2004, the preschool where I was working as a preschool teacher was certified across all levels. Firstly, bronze included aspects of electricity and heating, ventilation, water, food, materials, recycling, and cleaning chemicals; secondly, silver was about engaging children in activities focused on issues of sustainable development including social, economic, and environmental dimensions; and finally, gold included connecting the preschool's ESD activities with parents, the local community, and sharing activities locally, regionally, and with the global world.

In 1998, at the World Organisation for Early Childhood Education (OMEP) conference in Copenhagen, Denmark, I presented my work about ESD in ECE for the first time. At this time, it was obvious that there were few scientific national or international studies about ESD in preschool education. This led to my 2003 Masters of Education degree in which I focused on preschool teachers' practices and understandings about nature and the environment as relevant to the Swedish national curriculum.

After 1998, I was often asked to lecture about ESD. My engagement in the OMEP Swedish chapter and continuing participation led to my development of a wide network of researchers, teacher educators, and practitioners from all over the world. In 2007, UNESCO and the University of Gothenburg arranged the first international conference 'The role of early childhood education for a sustainable society' and it was at this conference that I first met Julie Davis. Later, at the first Transnational Dialogues in ECEfS Research (TND) gathering in 2010, I also met Sue Elliott. Since then, Julie, Sue, and I have collaborated on numerous publications, conference presentations and engaged in professional exchanges. In 2004, I started work at Mälardalen University in the Division of Early Childhood Education, lecturing in the preschool teacher program. Between 2016 and 2019, I was Head of the Division of Early Childhood Education.

My doctoral thesis completed in 2013 was entitled *Committed to the good of the world? Education for sustainability in the Swedish preschool.* This was the first Swedish thesis on education for sustainability in Swedish preschools. I investigated teachers' knowledge about ECEfS and scrutinised the EfS ambiguities, complexities, and contradictions within Swedish preschools. I also analysed children's agency and meaning-making and explored how these were manifested in EfS using a theoretical framework guided by critical theory and an ecofeminist perspective.

In 2018, Sue and I initiated the European Early Childhood Education Research Association (EECERA) Special Interest Group (SIG), *Sustainability in Early Childhood Education.* Now as a professor in early childhood education at Mälardalen University, Sweden, I lead the research cluster *Children and young people in school and in society* (BUSS) and the research groups: *Childhood in the Anthropocene–Education and sustainability* and *Space and Place in Early Childhood Education and School.* I am grateful to be part of this journey of developing knowledge about ECEfS and hope this book can inform and inspire further research and support early childhood educators all over the world.

Explaining the Text

Early Childhood Education for Sustainability (ECEfS) has been an emerging field for about 30 years. As three pioneers, with extensive backgrounds in ECE, EC teacher education and who are also researchers in this field, we believe we have unique insights into ECEfS. This is especially so regarding its antecedents—a field does not just emerge from nowhere—and what has influenced it—people, ideas, movements, shifting policies, politics, and actual events. Our participation in growing the field—first from our own Australian and Swedish perspectives, and more recently as participants in an expanding international and global movement, we feel we have been in powerful positions to understand background influences, see emerging trends, and interweave these many bits and pieces into, a sort of narrative.

We acknowledge, however, that there are indeed, multiple histories and stories but in the end, we can only tell our own story. It will be up to others to tell, dispute, and shape their own versions of how and when ECEfS started, how it has transpired, and where it might be in future.

Nevertheless, regardless of how it is understood and explained, we fervently believe that ECEfS has a key role in preparing young people, their families and ECE communities for contributing to the creation of a more sustainable world.

Brisbane, Australia — Julie M. Davis
Armidale, Australia — Sue Elliott
Västerås, Sweden — Eva Ärlemalm-Hagsér

References

Ärlemalm-Hagsér, E. (2013). *Engagerade för världens bästa? Lärande för hållbarhet i förskolan [Committed to the good of the world? Education for sustainability in the Swedish preschool]* (Doctoral thesis, Gothenburg Studies in Educational Sciences 335). Acta Universitatis Gothoburgensis.

Davis, J. (2010, 2015). *Young children and the environment: Early education for sustainability,* Cambridge University Press.

Davis, J., & Elliott, S. (Eds.) (2024). *Young children and the environment: Early education for sustainability* (3rd ed.). Cambridge University Press.

Elliott, S., Ärlemalm-Hagsér, E., & Davis, J. (Eds.) (2020). *Researching early childhood education for sustainability: Challenging assumptions and orthodoxies (Vol. 2).* Routledge.

Acknowledgements

The drafting of Chap. 4 would not have been possible without the generous contributions of colleagues with expertise and wisdom who have shared our journey over several decades. We warmly thank them for their international insights and commitment.

Associate Professor Ingrid Engdahl (Sweden)
Michelle Hocking (Australia)
Professor Michiko Inoue (Japan)
Professor Okjong Ji (South Korea)
Professor Minyi Le (China)
Associate Professor Nadine McCrea (Australia)
Glynne Mackey (Aotearoa/New Zealand)
Professor Ingrid Pramling Samuelsson (Sweden)
Associate Professor Jenny Ritchie (Aotearoa/New Zealand)
Dr. Ruth Wilson (USA)

Contents

Abbreviations

AAEE	Australian Association for Environmental Education
CoP	UN Conference of Parties
ECE	Early Childhood Education
ECEfS	Early Childhood Education for Sustainability
ECET	Early Childhood Educators Tasmania–South
EE	Environmental Education
EECERA SIG	European Early Childhood Education Research Association Special Interest Group
EEEC Vic. Inc.	Environmental Education in Early Childhood Victoria Inc.
EESSA	Early Education for Sustainability South Australia Inc.
EfS	Education for Sustainability
ESD	Education for Sustainable Development
FNP	First Nations Peoples
IPCC	Intergovernmental Panel on Climate Change
ITE	Initial Teacher Education
LGSWA	Little Green Steps Western Australia
NSW EYNC	New South Wales Early Years Nature Connections
OMEP	World Organisation for Early Childhood
PL	Professional Learning
QECSN	Queensland Early Childhood Sustainability Network
SD	Sustainable Development
SDGs	UN Sustainable Development Goals
The Alliance	The Australian Early Childhood Sustainability Alliance
TND	Transnational Dialogues in ECEfS Research
UN	United Nations
UNCRC	UN Convention on the Rights of the Child
UNICEF	United Nations Children's Fund
WCED	World Commission on Environment and Development
WSA	Whole School Approach

List of Figures

List of Tables

Chapter 1
Introducing Our Key Ideas and Perspectives

Good planets are hard to find (*Kickbusch*, 1999)

Abstract In this introduction, we provide an overview of our rationale for writing this booklet. As we are coming to the end (somewhat!) of our academic careers, we felt our perspectives on why, when, and how ECEfS has developed—in collaboration with local, national, and international colleagues—is a 'good news' story worth telling. We also take the opportunity to explain key terms related to sustainability, education for sustainability, and its relevance to early childhood education, setting up the contexts for the following chapters.

1.1 Introduction

This slim volume comes at a time when there is a convergence of global crises—climate disruptions, the COVID-19 pandemic, geopolitical tensions in Europe, the Middle East, and the Asia–Pacific—overlayed by ongoing and equally disruptive issues of biodiversity loss, cumulative plastic pollution, increasing poverty and wealth disparities and refugee crises, including climate refugees. Our perspective on these crises is from our shared standpoint as early childhood teacher educators, researchers, and advocates. We have each been at the forefront of nurturing the field of Early Childhood Education for Sustainability (ECEfS) for more than 30 years (over a collective 100 years between us!) from its nascent beginnings to where ECEfS is now a recognised field of learning, teaching, and enquiry within broader spheres beyond early childhood education.

As colleagues and co-authors, we bring to this volume our different EfS experiences over the decades. As our profiles illustrate, we have each brought distinct backgrounds to our respective journeys. Julie and Sue grew up in Australia—Julie in the tropical north; Sue in cooler southern Australia, while Eva grew up in Sweden, closer to the Arctic Circle. We each have had different experiences in early childhood education (ECE), worked in different kinds of universities, and had different opportunities and connections with both ECE and EfS. Julie's early career was as a primary school teacher, working and living in remote Australian First Nations communities,

J. M. Davis et al., *Early Childhood Education for Sustainability*,
SpringerBriefs in Education, https://doi.org/10.1007/978-981-97-8405-9_1

then studies in environmental issues, a Master of Environmental Education and later a doctorate, before becoming deeply involved for ECE. Sue began her professional career teaching in early childhood education, diverged into science degree studies, and then effectively combined these strands in a doctorate with a focus on ECEfS. Eva started her professional career as preschool teacher in Sweden, followed by studies in Educational Science at a Masters level, then a Doctor of Philosophy. Thus, we have brought together different lenses to this 'project' of ECEfS, and how we now collectively write this history. We also recognise many others around the globe who have been instrumental in this history. A few, from across Asia, the Pacific, and Scandinavia, we invited to share their thoughts and recollections (see Chap. 5); we sincerely thank them for so generously contributing to this text.

Nevertheless, while we have come separately to what are now some quite shared viewpoints, we have also had the great fortune to meet both in person and online in various ways over the years and have been able to drive forward together—making and shaping as well as sharing our own reflections and critical re-examination of ECEfS over time. These latter shared experiences have enabled us to shed earlier feelings of isolation and marginalisation; our collaborations have had powerful lasting effects in driving our research and advocacy internationally.

We are not getting any younger—two of us are already retired!—and we know that we each hold personal knowledge and perspectives—much shared, but some quite personal—that has informed who we are, why we have been motivated to practice and research in this space, what and how the field of ECEfS has emerged. With Julie at the helm, we all agreed it was time to put down in words how we have seen the field evolve and present our understandings, theories, strategies, and serendipities while we still have the desire and ability to do so. In an overt way, we see this book as our legacy, to be used, added too, and critiqued by future educators and scholars with an interest in ECEfS.

1.2 How We Wrote This Book

The idea of writing this history had been brewing for some time. While we have each written accounts of the emerging ECEfS field over the years, it has only been in the past 5 years or so that we have thought about pulling the multiple strands of our work together into a more cohesive historical story and speculations about the future. This has been driven largely by the fact that we are more or less at the end of our careers. At some point, we really do need to hang up our keyboards!

To advance this writing project, we collectively agreed on a structure and then bid to lead chapters and topics, an easy process as we were aware of our own interests and strengths and had developed strong trust in each other's thinking and writing. Therefore, while we consider ourselves equal co-authors, each chapter had a designated lead while the other authors offered editing and content suggestions and critique. Thus, this book demonstrates our own individual writing styles and approaches to content. It is not a chapter book where each chapter has the same formula—it's more a bricolage or a collection—created from diverse ways of thinking and writing that reflect our unique personalities and histories. There is some introspection (Preface), informality (Introduction), historical reflection (Chap. 2), and academic framing (Chaps. 3 and 4). Personal reflections are throughout, allied with commentaries from some of our international colleagues (Chap. 5) and there is speculative, future-gazing in Chap. 6. We rather like what has emerged and hope that novices to ECEfS, as well as longer-standing practitioners, researchers, and policymakers, will find the text an interesting read.

1.3 Where to Start?

We see the climate crisis as a children's crisis, and we are not alone in this view. We know that young children and future generations have been identified—by multiple authors from diverse disciplines—as the most vulnerable to ongoing climate crises and unsustainability. It is already known that today's babies will live/are living through three to four times as many extreme climate events as their grandparents (IPCC, 2023), and they will need to emit eight times less CO_2 over the course of their lifetimes than their grandparents (Thiery et al., 2021). Other commentators with much bigger platforms than we have express that climate change is the biggest challenge facing the planet. The UN Secretary-General António Guterres, in his concluding remarks at 2022 Conference of Parties (CoP 27) stated: '*Our planet is still in the emergency room... We need all hands on deck to drive justice and ambition. The world still needs a giant leap on climate ambition ... Together, let's not relent in the fight for climate justice and climate ambition*'.

However, while the climate crisis is triggering current attention and a growing sense of urgency among many authors, commentators, educators, and researchers, our interest is much wider than climate change and goes back much further. As our individual profiles demonstrate, we have all focused on matters of nature, environment, social justice, and sustainability—and their impacts for children now and in future—over several decades. For example, in 1996, Julie wrote an 'eco-bulletin' to her university colleagues in the School of Early Childhood as an early response to her attendance at an international conference, where David Suzuki was a keynote presenter. In this bulletin, she articulated her growing concerns about the state of the global environment, and its impact on current and future generations, such thinking was uncommon on the educational agenda in EC teacher education courses at the time. In these early days too, terms, such as 'greenhouse effect' and 'global warming', were just becoming recognised, well before 'climate change' and 'climate catastrophes' entered everyday vocabularies. Also, during this time, ozone layer concerns had come and almost gone, offering some hope for the Earth's remediation.

Given that the early years, from birth to eight, is the most critical phase for learning and development in the human life span, we consider that ECE must prioritise education for sustainability (EfS)/education for sustainable development (ESD). It must be the foundation stone for thinking, pedagogies, and practices with the explicit intent of addressing present and future challenges for children and young people. Again, we are not alone in this thinking, multiple authors and commentators have identified ECE as the starting point for engaging with learning about and for sustainability (Hedefalk et al, 2014; Pramling Samuelsson & Kaga, 2008; Somerville & Williams, 2015; The International Commission on the Futures of Education, 2021; UNESCO, 2014).

Over the years, education has been seen as an important driver for learning about and for sustainability, but how this is achieved has been discussed and problematised for decades, see for example Jickling's (1992) article, *Why I don't want my children to be educated for sustainable development.* He argued that essential philosophical questions as well as moral and ethical dilemmas involved within sustainability were generally overlooked in education. In contrast, Sterling (2004) saw great benefits in educating for sustainability and argued that it provided a unique opportunity to reform all education. Amidst these debates and since different international interpretations as to how education and sustainability fit together have appeared in various research and policy documents. For example, environmental education (EE), education for sustainable development (ESD), education for sustainability (EfS), and environmental sustainability education (ESE) are terms quite widely employed, while recently climate change education (CCE) and disaster resilience education (DRE) have become evident. Much has already been written on these constructs and their meanings—by many authors and researchers as well as us. Here we give a potted account of these varied constructs and how they relate to ECE. To further assist your understanding as you read, please refer to 'Key Terms' at the back of this booklet.

1.4 Sustainability/Sustainable Development

While the emergence of ECEfS as a field of education practice and enquiry is quite recent (no more than about 30 years), it is linked to longer-term concerns arising from the 'environmental crises' of the 1960s. World population growth (notably in poorer non-Western nations combined with excessive consumption of resources (notably in richer Western nations)—dictated by adherence to growth economics—were identified as key underpinning causes of environmental degradation and social injustice. This was first raised in influential books and reports (Ehrlich, 1968; Lovelock, 1979) that emphasised excessive human behaviour—especially at a global level—as the driver of these multiple 'environmental' crises, especially in the West. Global social, economic, and educational change was called for.

Arguably, one of the key starting points for global understandings of the nexus between humans and environment was the United Nations Conference on the Human Environment held in Stockholm, Sweden, in 1972. This meeting was the first world conference that placed environmental concerns at the core of human development discussions. The establishment of the United Nations Environment Program (UNEP) and World Environment Day were outcomes of this meeting. The term 'sustainability', as a concept, appeared in a 1972 text 'Blueprint for Survival' (Goldsmith &

Allen, 1972), a novel idea about no-growth economies, at a time when the global environmental movement was gaining traction. Books such as *Silent Spring* (Carson, 1962), *The Population Bomb* (Ehrlich, 1968), *The Limits to Growth* (Meadows et al., 1972), *Small is Beautiful* (Schumacher, 1973), and *Gaia* (Lovelock, 1979) echoed the mood for change reflecting an ecological discourse that maintained human lifestyles including world population growth, materials consumption, industrialisation, pollution, overcropping, and resource depletion were having severe negative impacts for the natural environment on which humans entirely depend. Inequalities between richer global minority nations and poorer global majority nations were also prime ideas reflected in these discourses. The clear message was that humans were seriously damaging a fragile Earth ('Good planets are hard to find', Kickbusch, 1999) at rates that could not be sustained and were rapidly becoming irreversible for future generations.

In the 1980s, the term 'sustainable development' (SD) became more commonplace, signalling a shift from articulating the problems of the human–environment nexus, to determining a 'way out'. This was a call for sustainable development, rather than the exploitative growth and economic development 'at all costs' which was already so entrenched, especially in global organisations such as the World Trade Organisation and the World Bank.

The most common definition of SD, still used extensively today, is that stated in the *World* Commission on Environment and Development (WCED) *Brundtland Report, Our Common future* (1987), 'development that is able to "meet" the needs of the present without compromising the ability of future generations to meet their needs' (WCED, 1987, p. 393). While remaining an important reference point, how SD is interpreted and described internationally is a topic of ongoing debate (Hopkins, 2009; Sauvé, 2005). Some argue that 'sustainable' and 'development' are oppositional, and the key global challenge is to find development paths within planetary and social boundaries for collective human well-being as illustrated by Raworth's (2017) circular or doughnut economics, rather than centring on linear and increasing economic development. Thus, an alternative interpretation of 'sustainability' removes reference to 'development', referring to sustainability as multidimensional comprising interlinked political, economic, social, cultural, and natural dimensions (UNESCO, 2021). In Australia, for example, it is more common to use 'sustainability', while in Europe SD is typically employed. Nevertheless, we and others working in sustainability, subscribe to multidimensional understandings and recognise that this must underpin all decision-making for realising equitable global futures.

Common to virtually all viewpoints is the understanding that sustainability must address social and economic issues as well as environmental matters. However, even these ideas lie on a spectrum, with some definitions perceived as 'weak' and some considered 'strong' (Ruhs & Jones, 2016). Weak sustainability can be considered as development that only scratches the surface of the necessary social and economic changes. Green business models that promote sustainable practices but continue to expand consumption of goods and services fall into this category. Essentially, weak definitions of SD continue to emphasise **anthropocentric worldviews**. Strong sustainability, on the other hand, embraces **ecocentric worldviews** and the intrinsic valuing and preservation of nature, the interdependencies of social, economic, and environmental dimensions, and equitable social and economic development. Calls to ensure that non-human lifeforms and Earth elements such as the atmosphere, water, and soils must be considered in discussions alongside and equal to human well-being and advancement.

The most recent global restatement of ideas reflecting sustainability can be seen in the Sustainable Development Goals (SDGs) as integral to UNESCO's Agenda 30. These offer a broad framework comprising seventeen interrelated goals for global implementation of sustainability initiatives for change, particularly across social and economic spheres. Critique and implementation of the SDGs in ECE settings has already begun with current research suggesting the SDG's have the potential to be an important tool for rethinking how to enact ECEfS (Carr et al., 2021; Sonter et al., 2024). Anecdotally, uptake of the SDGs in ECE seems more evident internationally than the previous UNESCO global initiative, the Decade of Education for Sustainable Development 2004–2015 (UNESCO, 2005). However, we noted at the time that momentum for change was on the rise over the Decade such that ECE was beginning to be recognised by policymakers as critical to integrating sustainable development across education (UNESCO, 2014). Since then, rising concerns about climate catastrophes probably also account for engagement with the SDGs and how thcy might be helpful to the ECE field in contemporary times. Broadly, the fields of EfS and ESD continue to be associated with the discourses emanating from these earlier decades, influenced by, and related to, the underpinning constructs of sustainability/SD.

1.5 The Anthropocene/Ecocene

Strongly associated with sustainability/SD is the concept of the **Anthropocene.** This is a relatively recent concept coined in the early 1980s by ecologist Eugene Stoermer and popularised by Nobel Prize-winning atmospheric chemist, Paul Crutzen (Steffen et al., 2007, 2018). Not quite an official epoch in the sense of the Pleistocene or the Palaeolithic Age, it refers to the period beginning with the Industrial Revolution in the mid-1770s. The influence of the human species became so significant that it dramatically reshaped the Earth and its ecological systems, but not for the better. It is argued that the Anthropocene highlights negative human global impacts and the challenges of living within the Earth's boundaries, and that we exceed these boundaries at our peril. The climate change crisis clearly illustrates that humans have already transgressed six of nine **planetary boundaries** (Richardson et al., 2023), strengthening arguments for urgent change across social, economic, political, and environmental dimensions.

Several researchers are making connections between anthropocentric forms of human development, global sustainability, and COVID-19 (Scholes & Scholes, 2020). *The Lancet's* 2020 report (Watts et al., 2021) describe a convergence of crises with multipronged solutions required. We have witnessed international urgency, increasing scientific literacy, and resolution to address the pandemic and we argue a similar mission is required to address climate change. The global community has well demonstrated its capacities for rapid and often contentious change in how we can live and work during a pandemic; thus, we contend that we must do this again to avoid the ever-widening impacts and inequities of climate change for all species now and into the future.

The **Ecocene** offers a way forward, proposed as a new epoch to work towards by Boehnert (2018). The Ecocene is described as an ecological epoch requiring shifts in worldviews and systems to promote ecologically informed living. We—and many others who remain optimistic about human capacity to change—believe that personal and collective global transformation *is* feasible; indeed, we have been working in ECEfS because it offers many positive responses to the issues and dilemmas caused by (un)sustainability. It is our goal that all EC educators and their communities with commitments to young children identify this as core business for children's healthy and sustainable futures.

1.6 Environmental Education, Education for Sustainability, Education for Sustainable Development

As with its profiling of growing global human–environment concerns, the 1972 Stockholm Conference was also the first global meeting to recognise a key role for education in ameliorating environmental harms, and for contributing towards more healthy, just, and sustainable relationships between people and planet. As with variations in terminologies around sustainability and SD, there are also international variations across the education field with commonly employed terms that create some confusion. The Brundtland Report (WCED, 1987) identified the key role of education in enacting sustainability; hence, this report strongly influenced the emerging field of EfS/ESD. However, this advance did not develop in a void. Its antecedent is EE, a field of educational practice and inquiry drew on the somewhat alarmist contexts of the 1960s and 1970s previously described (see Chap. 2). EE emerged as the key remedy for mounting environmental issues and was promoted in the Western world through pivotal international UNESCO meetings, such as The Belgrade Charter (UNESCO UNEP, 1975) and The Tbilisi Declaration (UNESCO UNEP, 1977). There is a certain irony that what began over 50 years ago to shift anthropocentric thinking through education remains a somewhat slow work in progress, especially as calls to address the state of the planet have become more deafening.

Environmental Education (EE) has long predominated and retains currency in Japan and the USA. More recently, however, a shift towards ESD and EfS has occurred while newer variations such as environmental sustainability education (ESE) have surfaced. While not quite the same, but related, climate change education (CCE) and disaster resilience education (DRE) have also surfaced. Mostly, European countries and specifically UNESCO refer to ESD, while Australasia employs

EfS. While such varied terminology can be frustrating, we believe that the sentiment embodied in the different terms is more important than the actual terminology used. Whichever term is applied in ECE, fundamentally across all aspects of early childhood services 'We must reorient our approach from solving discrete siloed problems to navigating multidimensional, interconnected and increasingly universal predicaments' (United Nations Development Program (UNDP), 2020, p. 5).

For many, EE has traditionally focused more on being in, and learning from, the natural environment, with emphasis on understanding, preserving, and protecting ecological systems. In ECE, this is referred to as nature play or nature education which has a long history in ECE (Ärlemalm-Hagsér & Elliott, 2021). Generally speaking, EE has been the province of science and/or geographical education and conducted mostly in schools and higher education institutions where 'discipline subjects' prevail. However, we consider that one of the key distinctions between EE and contemporary EfS, regardless of educational setting, is the empowerment focus on individuals and communities to take action for addressing and transforming current environmental/ecological and sustainability issues in multidimensional ways, thus creating healthy, just and sustainable futures for all (see Chap. 2).

Additionally, in going beyond environmental/ecological systems learning and thinking, EfS is distinguished by recognising a much wider understanding of systems and applying systems thinking to create change. Drawing on ecological and social systems theory, EfS is about deeply understanding the operation of interconnections and complexities in systems. Thus, it moves from more mechanistic, reductionist, and fragmented approaches to learning and change—often driven by focusing on science knowledge and individual behavioural change—to more complex, ecological, interdependent, interdisciplinary, and community-driven approaches (Sterling, 2004), that have greater impact. For example, addressing sustainability with children by installing a worm farm or adding a recycling bin is simply one fragment of much bigger systems that must be explored to collectively build whole centre/whole school thinking.

Systems thinking underpins whole centre/whole school approaches to sustainability such as sustainable schools and eco-kindergartens that seek to develop and implement cohesive, collaborative, and transformative approaches across all elements of educational systems—formal and informal curriculum, pedagogy, leadership, management services—both human and physical, and community connections.

1.7 Early Childhood Education for Sustainability

Inevitably, the practices, principles, and trajectory of EfS/ESD also underpin ECEfS; however, we see ECEfS as quite different from EfS—a distinct subfield of education—because of their different historical, theoretical, and pedagogical underpinnings. (This is dealt with in more detail in Ch 2.) Like EfS/ESD, more broadly, ECEfS supports the multidimensionality of sustainability, and supports 'systems' thinking—such as 'whole of centre' approaches—to inspire change as well as the enactment of transformative, empowering, and participatory early education around sustainability issues, topics, and experiences. As a result, we hope to see the deepening development of 'cultures of sustainability' being generated and embedded in ECE settings through ECEfS.

We also see our brief for ECEfS as not being restricted; thus, it includes a broad range of contexts involving young children—homes, childcare centres, kindergartens, preschools, schools, environmental education centres, zoos, botanical gardens, Indigenous heritage sites, and many other community sites. However, we personally focus on more formally established educational sites such as kindergartens, preschools, childcare centres, and schools relevant to our professional histories, and because we recognise these sites as having capacities for leading deep and wide change across the ECE field.

We also reflect that we bring a distinct lens to ECEfS that has not necessarily been derived from the wider EfS field. This relates to our shared stance regarding children and childhood and as educators, we continue to be engaged with the EfS/ESD field and are cognizant of its shifts and changes over the past 30 plus years. In the 1990s, the ECE field began to move beyond deficit images of children by embracing theoretical movements such as the Sociology of Childhood that reconceptualised images of children as active social participants in the construction of their lives with voices to be heard, listened to, and acted upon. Allied with this shift, the UN Convention on the Rights of the Child (UNCRC) (UN, 1989) was adopted almost universally. These movements have been significant provocations for rethinking stances about children, with implications for ECEfS. Thus, we have long advocated for positive and empowered images of children as rights holders and social activists, a strong foundation for educators to build collective and systemic change across the dimensions of sustainability, critically beginning in early childhood. We argue these theoretical shifts along with child-centred play-based pedagogies (Elliott & Davis, 2009) offer unique advantages in ECE and have placed ECEfS well ahead of much of EE/EfS, at a time when more broadly the focus was on didactically teaching content.

Discussions and critiques about children's rights within ECEfS have deepened over time as we have sought to bring our ideas about children into deeper understandings about the challenges of climate change and sustainability. For example, lead author Davis (2014) challenged the anthropocentric focus of the UNCRC (1989), re-imagining rights within a broader worldview to include agentic, participatory, intergenerational, collective, Indigenous, and bio/ecocentric rights (Davis, 2014). Elliott and Davis (2018) have also applied a critical lens to the foundational theories of Bronfenbrenner, so highly revered in ECE. Similarly, recent publications by Visnijic-Jevtic et al. (2021) and Fenton-Glynn (2019) critically examine rights and justice in current global contexts. Notably in 2020, the UN Human Rights Council's 2020 added 'the right to a healthy environment' as integral to the Rights of the Child (OHCHR, 2020). Thus, significant rethinking has been embraced about children's rights in these increasingly uncertain times. We hope that early childhood educators will engage with and enact such rethinking as integral to implementing ECEfS.

1.8 Synergies Moving Forward

The following chapters in our historical account of ECEfS expand upon many of the points introduced above. Chapter 2 takes an historical and international perspective to discuss antecedents of contemporary ECEfS, highlighting events and triggers that have promoted its local and international evolution, as well as significant international drivers that have shaped and are continuing to shape ECEfS.

Chapter 3 traces changes in theories and pedagogies that have led to present constructions of ECEfS and turns attention to some of the newer theories such as post-humanism and new materialism that continue to shape and reshape the field.

In Chap. 4, we explore shifts in research that have occurred alongside theoretical and pedagogical change. These include our recent work in building international partnerships and research collaborations to continue the expansion and deepening of what we do and think. We highlight key research that showcases what has happened in ECEfS over the past 20 years and the sharing of ECEfS in wider national international fora.

In Chap. 5, we turn to identifying key enablers and turning points in ECEfS. Rather than simply drawing on our own ideas and perspectives, we do this alongside an analysis of interviews with ten key international players who have shaped and are shaping ECEfS, ranging from the 1990s to the present, across a diversity of nationalities. Through our analysis, we illustrate variety and depth in what we are saying about ECEfS in this short history.

Finally, in Chap. 6, we shift our gaze from the past to the future. We discuss ongoing challenges for the field, with particular reference to political, institutional, and policy inertia. We also summarise what we consider to be key learnings from our collective 100 years involved in and leading ECEfS. We focus on capacity-building networks as essential enablers in keeping us sane, connected, and motivated! The chapter and the book conclude with our take on what might be next for ECEfS and its ongoing trajectory. Certainly, the job is not done!

References

Ärlemalm-Hagsér, E., & Elliott, S. (2021). Outdoor play and learning in early childhood education. *Oxford Bibliographies*. https://doi.org/10.1093/OBO/9780199756810-0273

Boehnert, J. (2018). *Design, ecology, politics: Towards the Ecocene*. Bloomsbury.

Carr, V., Elliott, S., & Ärlemalm-Hagsér, E. (2021). Childhood Environmental Education—Sustainable development goals and early childhood environmental education: Awareness, understanding, and transformation. *Editorial Special Issue: International Journal of Early Childhood, 9*(1), 5–8. https://naturalstart.org/research/ijecee/volume-9-number-1

Carson, R. (1962). *Silent spring*. Houghton Mifflin Company.

Davis, J. (2014). Examining early childhood education through the lens of education for sustainability: Revisioning rights. In J. Davis & S. Elliott (Eds.), *Research in early childhood education for sustainability: International perspectives and provocations* (pp. 21–37). Routledge.

Ehrlich, P. (1968). *The population bomb*. Sierra Club/Ballantine Books.

Elliott, S., & Davis, J. (2009). Exploring the resistance: An Australian perspective on educating for sustainability in early childhood. *International Journal of Early Childhood, 41*(2), 65–77.

Elliott, S., & Davis, J. (2018). Challenging taken-for-granted ideas in early childhood education: A critique of Bronfenbrenner's ecological systems theory in the age of post-humanism. In K. Malone, E. Barratt-Hacking, & A. Cutter-Mackenzie (Eds.), *Research handbook on childhood-nature: Assemblages of childhood and nature research (Springer international handbooks of education)* (pp. 1–36). Springer.

Fenton-Glynn, C. (Ed.). (2019). *Children's rights and sustainable development: Interpreting the UNCRC for future generations.* Cambridge University Press.

Goldsmith, E., & Allen, R. (1972). *Blueprint for survival.* Ecosystems Ltd.

Hedefalk, M., Almqvist, J., & Östman, L. (2014). Education for sustainable development in early childhood education: A review of the research literature. *Environmental Education Research, 21*(7), 975–990. https://doi.org/10.1080/13504622.2014.971716

Hopkins, C. (2009). Enough, for all, forever: The quest for a sustainable future. *Education Canada, 49*(4), 42–46.

IPCC UNEP. (2023). Climate change 2023: Synthesis Report | UNEP-UN Environment Programme.

Jickling, B. (1992). Why I don't want my children to be educated for sustainable development. *Journal of Environmental Education, 23*(4), 5–8.

Kickbusch, I. (1999). *Good planets are hard to find.* World Health Organisation (WHO) Healthy Cities Papers No. 5. https://www.ilonakickbusch.com/kickbusch-wAssets/docs/Good-planets-are-hard-to-find.pdf

Lovelock, J. (1979). *Gaia: A new look at life on Earth.* Oxford University Press.

Meadows, D. et al. (1972). *The limits to growth.* Potomac Associates—Universe Books.

Office of the High Commissioner for Human Rights (OHCHR). (2020). *Realizing the rights of the child through a healthy environment.* https://www.ohchr.org/EN/NewsEvents/Pages/DisplayNews.aspx?NewsID=26022&LangID=E

Pramling Samuelsson, I., & Kaga, Y. (Eds.). (2008). *Early childhood and its contribution to a sustainable society.* UNESCO.

Raworth, K. (2017). A Doughnut for the Anthropocene: humanity's compass in the 21st century. *The Lancet Planetary Health, 1*(2), 48–49.

Richardson, K., Steffen, W., Lucht, W., Bendtsen, J., Cornell, S. E., Donges, J. F., et al.(2023). Earth beyond six of nine planetary boundaries. *Science Advances, 9*, 37.

Ruhs, N., & Jones, A. (2016). The implementation of Earth jurisprudence through substantive constitutional rights of nature. *Sustainability, 8*(2), 174. https://doi.org/10.3390/su8020174

Sauvé, L. (2005). Currents in environmental education: Mapping a complex and evolving pedagogical field. *Canadian Journal of Environmental Education, 10*(1), 11–37.

Scholes, M., & Scholes, R. (2020, June 10). Climate change, biodiversity loss and other global ills share root causes. *The Conversation.* https://theconversation.com/climate-change-biodiversity-loss-and-other-global-ills-share-root-causes-140183

Steffen, W., Crutzen, P. J., & McNeill, J. R. (2007). The anthropocene: Are humans now overwhelming the great forces of nature? *Ambio, 36*(8), 614–621.

Steffen, W., et al. (2018). Trajectories of the Earth system in the anthropocene. *PNAS, 33*(115), 8252–8259.

Schumacher, E. F. (1973). *Small is beautiful: A study of economics as if people mattered.* Harper & Row.

Sterling, S. (2004). Higher education, sustainability and the role of systemic Learning. In P. B. Corcoran & A. E. J. Wals (Eds.), *Higher education and the challenge of sustainability: Problematics, promise and practice* (pp. 49–70). Kluwer Academic.

Somerville, M., & Williams, C. (2015). Sustainability education in early childhood: An updated review of research in the field. *Contemporary Issues in Early Childhood, 16*(2), 102–117.

Sonter, L. J., Kemp, S., Lamb, C., & Willey, S. (2024). Exploring the sustainable development goals: beyond the environmental dimension. In J. Davis & S. Elliott (Eds.), *Young children and the environment* (3rd ed.). Cambridge.

The International Commission on the Futures of Education (2021). *Futures of education: Learning to become, Progress update*. UNESCO.
Thiery, W., Lange, S., Rogel, J., et al. (2021). Intergenerational inequities in exposure to climate extremes: Young generations are severely threatened by climate change. *Science, 374*(6564), 158–160.
Visnijic-Jevtic, A., Sadownik, A., & Engdahl, E. (Eds.). (2021). *Young children in the world and their rights: Thirty years with the United Nations Convention on the Rights of the Child*. Springer Nature.
UNESCO UNEP. (1975). *The Belgrade Charter: A framework for environmental education*. http://unesdoc.unesco.org/images/0001/000177/017772eb.pdf
UNESCO UNEP. (1977). *Intergovernmental conference on environmental education: Tbilisi (USSR), 14–26 October 1977. Final Report*. UNESCO.
United Nations (UN). (2022). Statement by the Secretary-General at the conclusion of COP27 in Sharm el-Sheikh, United Nations Secretary-General.
United Nations (UN). (1989). *United Nations Convention on the Rights of the Child* (UNCRC). www.ohchr.org/en/professionalinterest/pages/crc.aspx
UNDP. (2020). *The next frontier: Human development and the Anthropocene*. UNDP.
United Nations Educational, Scientific and Cultural Organisation (UNESCO). (2005). Decade of education for sustainable development 2005–2014 .*Draft International Implementation Scheme*. UNESCO.
United Nations Educational, Scientific and Cultural Organisation (UNESCO). (2014). *Shaping the future we want: UN Decade of Education for Sustainable Development (2005–2014) final report*. UNESCO.
United Nations Educational, Scientific and Cultural Organisation (UNESCO). (2021). *Re-imagining our futures together: A new social contract for education*. UNESCO.
Watts, N., et al. (2021). Countdown on health and climate change: Responding to converging crises. *Lancet, 397*(10269), 129–170. https://doi.org/10.1016/S0140-6736(20)32290-X
World Commission on Environment & Development (WCED). (1987). *The Brundtland report: Our common future*. Oxford University Press.

Chapter 2
ECEfS/ESD: Historical National and International Antecedents

Abstract In this chapter, we chart the history of early childhood education for sustainability (ECEfS) with a focus on early beginnings in Australia and Sweden. We then explore the international momentum developed over time through research and policy which has led to a now consolidated field of ECEfS research and practice. Finally, we alert readers to our reflections about the strengths and tensions experienced along the way and offer an almost four-decade chronology for those seeking further insights.

2.1 Introduction

This chapter charts the historical antecedents for early childhood education for sustainability (ECEfS) as a recognised field of research and practice. We explore localised, national, and international 'ripples of green' that have propelled us forward, first beginning in the 1980s. Over time, momentum has built such that we are now well-positioned to write about the history of this field. We first orient the reader with early beginnings in both Australia and Sweden, then consider their respective histories of networking for organisational change through initiatives, research, publications, and policies. We then consider international landscapes, with a focus on history, research, and policy, and conclude by reflecting on strengths and tensions over time. Lastly, we draw on a previously published chronology (Davis & Elliott, 2014) and more recent international perspectives (Davis & Elliott, 2024; Elliott et al., 2020) to offer an updated chronology comprising some significant publications and events over an almost four-decade ECEfS/ESD history.

J. M. Davis et al., *Early Childhood Education for Sustainability*,
SpringerBriefs in Education, https://doi.org/10.1007/978-981-97-8405-9_2

2.2 Early Beginnings—Australia and Sweden

Here, we outline our familiar starting points—Australian and Swedish contexts for ECEfS/ESD. Again, we reiterate that we can only speak from our own perspectives—we know that others in these countries and more distant countries will have their own unique starting points (see Chap. 5). We also highlight that as Australian and Swedish authors and advocates we were also reading and being shaped by international works simultaneously—we did not live in our respective national bubbles, even though access to papers and reports was a more difficult and slower process, before-the internet.

2.2.1 Australia: Pragmatic Beginnings

Initially, in the 1980s and 1990s the focus in Australian ECE contexts was very much pragmatic; committed individual educators were aware of environmental issues and actively seeking ways to implement EE with young children. Initially, there was a confluence of practitioner publications in the 1980s and 1990s that pushed the boundaries beyond children's development and settings to more expansive environmental concerns with impacts for children and future generations. Two Australian publications captured this early focus, *Snails Live in Houses Too: Environmental Education for the Early Years* (Elliott & Emmett, 1991/1997) and *Playing for keeps: Environmentally responsible practices for early childhood learning* (Gordon Community Children's Centre Technical College, 1993). The sustainability practices ranged from the use of natural or repurposed play materials to productive gardening and environmentally responsible housekeeping. Even then, the authors were recommending holistic approaches to EE, rather than disconnected activities; and a whole service pedagogical approach is equally relevant today.

A feature of these publications was that they enshrined contemporary images of children, views that were disrupting older psychology-derived ideas of children—coming from the Sociology of Childhood field—as empowered and capable of taking environmental action. They ultimately made a strong case for the relevance of EE to the goals and programs of ECE. This was a move away from developmentally driven images of children as not yet capable, which had dominated ECE in Australia and elsewhere for many years (Elliott & Davis, 2009). Interestingly, children are perceived as proactive citizens aligned with the *United Nations Convention on the Rights of the Child* (UNCRC, 1989), plus the reconceptualised images of children as societal participants (James & Prout, 1990). Juxtaposed with these image shifts was the *Brundtland Report Our Common World* (WCED, 1987) which argued for intergenerational equity and heralded education as the way forward for sustainable global transformation. There could be no greater cause for intergenerational equity than the youngest citizens. On reflection, there seemed be a coalescing of thinking about children, their voices, and potential roles in creating sustainable futures. In many ways, the early days of ECEfS/ESD were at the forefront of adopting reconceptualised images of children, and we believe this early orientation has stood the field in good stead today.

Across the Pacific Ocean, we connected with Ruth Wilson in the United States of America who had begun publishing texts about ECEE, most notably *Fostering a Sense of Wonder During the Early Childhood Years* (1993) and *Environmental Education at the Early Childhood Level* (1994), plus many articles in the *Early Childhood Education Journal* (Wilson, 1995). From Australia, Davis (1998) was also an early contributor to this North American journal, thus building international links. These publications made a strong case for the compatible pairing of ECE and EE underpinned by a fundamental focus on children's experiences in natural environments. The latter publication (Wilson, 1994) offered practice case studies to illustrate effective ways forward. However, unlike other authors, Wilson understood the threats and challenges for children into the future from environmental destruction and introduced the concept of 'environmental justice'. Another early influential author was American Judith Dighe (1993), whose article 'Children and Earth' proposed a love, know, do approach—similar to the in, about, and for pedagogies—that were gaining momentum in EE. At this time, too, Australian academic Fien (1993) offered a critical theoretical lens to EE, emphasising education that developed critical thinking, political literacy, problem-solving, interdisciplinary approaches, and an environmental ethic, as defining characteristics. Drawing on these influences, Davis (1998) wrote a practitioner article combining the 'love, know, do' and 'in, about and for' ideas, to illustrate to ECE practitioners how concepts related to global and local environmental issues could be enacted through ECEE.

Early on, there was much attention towards tangible child-friendly actions such as reducing, reusing, and recycling whether lunch boxes or play materials were the foci. The pragmatic focus also extended to the impact of cleaning products and processes on human health, particularly for children (Immig, 2000). Efforts were made to investigate alternative and potentially less harmful ways of maintaining safe and hygienic ECE settings by Bridget Gardner, Michelle Hocking, and others. However, it was not until 2012 that the National Health and Medical Research Council (NHMRC) acknowledged 'some services may have a philosophy of environmental sustainability, and a consequent greater emphasis on considering the resources they use' (p. 2). This policy change took two decades of advocacy by those educators willing to challenge regulatory norms; and now, another decade on, further policy change is underway to more explicitly recognise sustainable practices. Thus, as mandated policies, these open the door for more educators to shift towards environmentally sustainable and healthy practices for all. Other topics of emerging interest to practitioners included active transport and energy conservation measures (Davis, 2005).

Unique to Australia, professional early childhood EE networks were established in the 1990s and most continue today. While a few mainstream early childhood professional organisations have progressed ECEfS nationally, such as the Australian Association for Environmental Education (AAEE) and the peak body Early Childhood Australia (ECA), the localised EE networks have been influential on the ground.

The networks varied in focus and operation, but initially, priorities for many educators were locating sustainable play resources, managing sustainable practices within policy such as nappy changing and cleaning, and being mutually supportive through regular professional gatherings. Such gatherings were critical at the time, as educators engaged in ECEE were most often perceived as marginal 'greenies', whose ideas and practices were not considered mainstream. The publication, *Early childhood environmental education: Making it mainstream* (Davis & Elliott, 2003) captured this sentiment. We note this publication drew largely on practice case studies in several Australian ECE settings. When updating the publication at ECA's request, we acknowledge theoretical framing or research studies to inform practice were limited; essentially, it was an advocacy piece. This republication has heightened our awareness about how far this field has travelled in two decades (Elliott & Davis, 2023).

On reflection, over the decades, environmentally committed early childhood educators—both practitioners and early childhood teacher educators—have needed to carve out their own spaces while straddling both the ECE and EE fields. This was not always easy; we were pushing boundaries on two fronts. Unfortunately, other professionals who perceived our mandate as marginal to the real work of ECE or EE frequently confronted us. Co-authors Davis and Elliott were often met with curiosity or alarm by many professionals across the education and environment sectors and aware of discussion in the broader community that denigrated this work. The perceptions we experienced were that young children were simply incapable of understanding or addressing environmental issues; even more so, environmental and sustainability issues were too dire to inflict on children at such a developmentally sensitive age (Elliott & Davis, 2009). Indicative was a seminar that Elliott attended, delivered by a prominent academic EfS guest speaker. She recalls advocating for ECEE during later discussion with the speaker, but his underwhelming response was to share a wallet photograph of his own young child to perhaps demonstrate some interest in young children. Similarly, Elliott participated in Australian Association for Environmental Education (AAEE) conferences from 1986 onwards and early childhood educators were very scarce. At the time, there was a strong focus on school-based environmental initiatives as worthwhile, but ECE mostly did not rate a mention. At the 1999 Sydney AAEE conference, co-author Davis and colleagues presented a symposium entitled 'Environmental education in early childhood: Mainstream not marginal'. We, and others, persisted in offering ECE-focused conference papers over many years; and, in 2003 with momentum growing, we successfully argued the case for an ECE SIG within the AAEE convened by co-author Elliott. One benefit of these often-dismissive exchanges was that it sharpened our arguments—by anticipating we were able to address concerns up front, and especially draw on research and other scholarly reports. For example, bringing ideas about young children's capabilities to the EE field, long driven by foundational scientific knowledges; and we think, not much attention to contemporary images of children as active participants with rights and ECE child-centred rather than school-based curriculum-centric pedagogies.

2.2.2 Sweden: Beginnings in the 1800s with Societal Change

A century earlier in the late 1800s, Swedish preschool pioneers (kindergarten teachers) instigated various forms of children's activities strongly inspired by Froebel's pedagogy and nature philosophy. At the time, Sweden was characterised by societal changes such as industrialisation, with families increasingly moving into urban areas to create cities. The loss of a Swedish agrarian society was keenly experienced, and nature was often perceived in opposition to urbanised industrial societies. Thus, the Swedish admiration for nature became an important cultural heritage to pass on to future generations (outdoor life or Friluftsliv remains integral to Swedish lifestyles). This inspired the Swedish kindergarten teachers (today the official professional title is preschool teacher) in urban kindergartens to organise excursions to forests and agricultural lands to develop children's respect for nature (Hatje, 1999). Nature was viewed as a place to be and learn and became integral to Swedish preschool practice. Another early Swedish preschool aim was to counteract the growing equity gaps in society and create greater harmony between different social classes.

2.3 Networking and Organisational Change in Australia and Sweden

In this section, we further address the local networking that has been central to promoting systemic organisational change through 'Communities of Practice'. Wenger (2006) defined these as 'groups of people who share a concern or a passion for something they do and learn how to do it better as they interact regularly' (p. 1). For us, such communities are typified by a shared interest in and commitment to ECEfS/ESD, a willingness to collaboratively learn together and collegially share ways of thinking and doing sustainability, whether practitioners, policymakers, service managers, or researchers. Outcomes can be seen in the various initiatives, research studies, publications, and policies recalled here. We trace the history of networking and organisational change evident in both Australia and Sweden, countries with long-standing active ECEfS/ESD leadership. Then, we document the broader international landscapes and networking including significant ECEfS/ESD research-focused groups.

2.3.1 Australian Historical Landscapes

Over the decades, 'Communities of Practice' were established by the Australian state-based professional educator networks, laying the groundwork for various initiatives, publications, and policies, all have contributed to progressing change and systemic

ECEfS uptake. To write a strictly linear historical account is challenging when dealing with many overlaying and intersecting systems over time.

Professional networks: As mentioned earlier, in Australia, state-based professional networks began with the establishment of Environmental Education in Early Childhood Vic. Inc. (EEEC Vic. Inc.) in 1992. This inspired other state-based networks in Queensland (QECSN established 1995), New South Wales (NSW ECEEN established 2003), and South Australia (EESSA established 2015). These networks have been highly effective in creating ECEfS 'Communities of Practice'; newcomers were welcomed, and regular gatherings offered informal and formal localised shared learning opportunities. Essentially these targeted ECEfS networks have been core to resourcing and creating change with practitioners and still do decades later.

The original Queensland EE network established in 1995, for example, had the following as its statement of purpose:

> *Environmental Education for Early Childhood*
>
> *is*
>
> *... about having vision,*
>
> *thinking for the future,*
>
> *creating change.....*
>
> *It's positive, practical.*
>
> *Education in, about and for the environment.*

In addition, several well-attended Australasian conferences were hosted: Christchurch, Aotearoa/New Zealand in 2006; Sydney, Australia in 2007; and, Melbourne, Australia in 2009 and over more recent years each network has hosted local conferences or events.

Most networks continue today, with some other affiliated organisations such as Little Green Steps Western Australia (LGSWA), Early Childhood Educators Tasmania—South and the Early Years Nature Connections—NSW (EYNC) now contributing. The networks share nationally as The Australian Alliance for Early Childhood Education for Sustainability (The Alliance) established at a summit of interested parties in 2016. The Alliance is called to action when there are significant national issues of common concern to address, such as the National Quality Framework review (Australian Productivity Commission, 2014) and the ECE curriculum policy review (Hadley et al., 2021). Historically, the AAEE and state chapters also offered networking opportunities through conferences and newsletters, and in particular, the ECE SIG. This group operated for some ten years from 2003, led by co-author Elliott, and raised the profile of ECE within the mainstream EE field.

Government initiatives: Local governments in Australia, more so than state or federal governments, were often proactive in promoting ECEfS, particularly in council managed ECE services guided by council operational sustainability targets. For example, the Gosford and Wyong Council in NSW instigated the Little Green Steps program in 2007 (Gosford & Wyong Council), while in Victoria, Hobson's Bay Council (2009) developed a Green Kinders' Tool Kit. At times multiple councils participated such as the Leading for Sustainability in Early Childhood and the More (Beyond) Little Green Steps projects which included workshops and mentoring (NSW ECEEN, 2013). On reflection, the networking around ECEfS across these various council initiatives promoted educator engagement; however, to what extent this was sustained post these periods of council activity is unclear.

State and federal government interest in ECE was briefly noted in several early policy documents such as *Learning to Care for our Environment: Victoria's Environmental Education Strategy* (Victorian Environmental Education Council, 1992) and *Today shapes tomorrow: Environmental Education for a sustainable future* (Commonwealth Department of Environment & Heritage, 1999). However, schools were more often the focus for targeted resourcing and funding and subject to political ebbs and flows. The most significant report to fully acknowledge the role of ECEE was a national EE review (Tilbury et al., 2005); serendipitously, the lead investigator had an ECE interest. Although a school sector review, this was the first national EE review inclusive of ECE with six pages by co-authors Davis and Elliott devoted to ECE. The report's introduction stated: 'In recent years, early childhood education in Australia has seen considerable development and now represents a growing area in EE. However, there is paucity of EE references, research and resources for these critical years' (Tilbury et al., 2005, p. 2). The 29 report recommendations were wide ranging across policy, research, and practice, but most importantly inclusive of ECE for the first time.

In the ECE field across the 2000s federal government policy shifts included a National Quality Framework (COAG, 2008) comprising both a National Quality Standard (NQS) (ACECQA, 2011) and, an inaugural, national curriculum policy (DEEWR, 2009). Both policies offered some incentive for the ECE field generally and leverage for those advocating for ECEfS. For example, the NQS Quality Area 3: Physical Environment included 'Standard 3.3: The service takes an active role in caring for its environment and contributes to a sustainable future' (ACECQA, 2011 [2013], 99–103). However, this was later watered down to a less explicit 'element' about environmental responsibility, after a significantly short-sighted review argued sustainability was not integral to quality and too challenging for educators to comprehend (Australian Productivity Commission, 2014). The curriculum policy *Belonging, Being and Becoming: Early Years Learning Framework for Australia (EYLF)* (Commonwealth DEEWR, 2009) encouraged educators to implement ECEfS through 'Learning Outcome 2: Children are connected with and contribute to their world' (DEEWR, 2009, p. 25) and the provision of natural outdoor playspaces (DEEWR, 2009, p. 16). These were valid policy starting points for fostering 'green' change, albeit not sufficiently potent or instructive for long-term ECEfS advocates. Comparative curriculum policy reviews by Ärlemalm-Hagsér and Davis (2014) and Weldemariam et al. (2017) attest to these concerns. An ongoing challenge with both the NQS and EYLF has been the need to view sustainability as embedded, across all seven NQS Quality Areas (ACECQA, 2018; Gaul & Watson, 2019) and all five EYLF Learning Outcomes (AGDE, 2022; DEEWR, 2009).

At state government level, two funded studies benchmarked the emergence of ECEfS, *Patches of green, early childhood environmental education in Australia: Scope, status and direction* (New South Wales Environmental Protection Agency, 2003) and *Examining environmental education in NSW early childhood education services: A literature review with findings from the field* (Elliott et al., 2016). The former highlighted the lack of systemic and multilevel approaches and attributed any uptake to committed 'green' individuals, the vision was to create a comprehensively green patchwork quilt. Just over a decade later, uptake was still varied and reportedly hindered by lack of educator confidence and knowledge around ECEfS and an ongoing need to fully embed sustainability across all ECE service pedagogy and practice (Elliott et al., 2016). At best, both state and federal recognition of ECE in promoting sustainable futures was ad hoc and limited. However, research and publications were building over this period and in some way addressed these multilevel government shortcomings.

Research and Research-based Publications: Beyond the early practitioner publications cited, Australian and neighbouring New Zealand research into ECEE/ECEfS began emerging in the 2000s. Most notable were studies by Davis (2005, 2009), Davis et al., (2008), Edwards and Cutter-Mackenzie (2011), Elliott and Davis (2009), and Vaealiki and Mackey (2008). Davis's (2009) review paper identified an ECEfS 'research hole' that became a clarion call for researchers. A subsequent increase in published studies was evident in later international reviews (Ardoin & Bowers, 2020; Hedefalk et al., 2014; Somerville & Williams, 2015). Building on this emerging researcher and practitioner uptake, Davis (2010) edited the first edition of *Young children and the environment: Early education for sustainability* with predominantly Australian chapter authors, but an eye to building international authorship and readership. This text was internationally the first to target a pre-service and in-service ECE teacher audience and has been translated into Korean (2014) and Chinese (2017). Second and third editions followed with an increasingly international authorship from across minority and majority countries (Davis, 2015; Davis & Elliott, 2024). In tandem, Australian practitioner publications increased and included both theoretical and pragmatic understandings about ECEfS (In chronological order Kinsella, 2007; Elliott et al., 2013; Young & Elliott, 2014; Sneddon & Pettit, 2016; Gaul & Watson, 2019; Early Childhood Australia, 2021; Elliott & Davis, 2023).

Professional organisations and curriculum policy initiatives: Peak mainstream bodies have been another avenue for promoting organisational change and ECEfS uptake. Early Childhood Australia facilitated a significant number of practitioner publications (Davis & Elliott, 2003; Early Childhood Australia (ECA), 2021; Elliott et al., 2013; Kinsella, 2007; Young & Elliott, 2014), professional learning through webinars, conferences, and online media, plus incorporated sustainability into their Professional Code of Ethics (ECA, 2016). However, generally sustainability action plans, policies, or position statements for leading the field remain less visible, despite attempts to raise sustainability as a key organisational policy initiative some years ago. In a state-based review by Elliott et al. (2016), few interviewees from various early childhood organisations and management bodies reported implementing sustainability-orientated philosophies, policies, or strategic plans, as essential to leading change for member services and educators. This remains an avenue to pursue for organisational change and leadership from the top.

Pre-service and In-service education: The approach to pre-service and in-service education has been described in the past as ad hoc, reliant on a handful of committed individuals in tertiary institutions and the Australian state-based networks (NSW EPA, 2003). Tertiary pre-service education includes universities offering a mandatory 4-year early childhood teaching degree, while vocational institutes (both public and private) offer shorter certificates and diplomas. Since 2012, vocational institutes have incorporated a sustainability unit in nationally mandated Children's Services Diploma and Certificate (TAFE NSW Training & Education Support Industry Skills Unit, 2010) courses, although anecdotally such units are being eroded. Some universities did introduce sustainability units into early childhood teaching degrees (Tilbury et al., 2005), however a review by Davis and Davis (2020) highlights ongoing concerns. Their study comprised an audit of EfS within university ECE pre-service courses based on publicly available course information, plus a systematic literature review. At a time, when many universities are purporting to subscribe to international, national, and institutional sustainability policies, they appear to be failing to translate these into teacher education courses, specifically ECE courses. A critical mass of new graduates with fuller understandings of sustainability principles and a pedagogical commitment is an important enabling factor and the tertiary sector must urgently address this gap. Evans et al. (2022) and Mulà and Tilbury (2023) offer international research insights about ways forward in this arena.

At Last, an Australian Curriculum Policy Shift! Building on these early decades, curriculum change became imminent during a 2021–2022 policy review. The previously noted Alliance made a submission, as a combined national voice of over one thousand members, to clarify misunderstandings about sustainability and support the adoption of a new sustainability principle in the revised *Belonging, Being and Becoming: The Early Years Learning Framework for Australia V2* (AGDE, 2022). The new sustainability principle now mandates educators to demonstrate sustainability in practice and describes sustainability as multidimensional including environmental, social, and economic dimensions, broadening the environmental focus of the 2009 version (DEEWR, 2009). We raise some concerns that the political dimension of sustainability (UNESCO, 2021) has been overlooked, but applaud the focus on children's active citizenship and agency. In addition, we argue that a more comprehensive approach across all five learning outcomes, not just the most obvious one about children connecting with the world, is required for effective implementation. However, after many years of advocacy we recognise this is pivotal in the history of Australian ECEfS and illustrates international leadership when only a handful of countries have adopted a sustainability principle in ECE curricula. The new principle aligns with the Sustainability Cross Curriculum Priority in the Australian Curriculum (ACARA, 2020), thus promoting the continuity of sustainability learning and teaching. Lastly, a second new principle embracing Aboriginal and Torres Strait Islander perspectives suggest synergies to explore alongside the sustainability principle. Co-authors

Davis and Elliott are basking at this pivotal moment in the knowledge that almost four decades of advocacy in Australia has at last been recognised and significant organisational change at many levels is now likely.

2.3.2 Swedish Historical Landscapes

From its foundations in the 1800s during a period of great social change and cultural valuing of nature, a modern history of preschool emerged in Sweden.

Beginnings of change: The modern Swedish preschool began in the 1960s driven by the necessity to grow the Swedish labour force with women perceived as essential in the labour market. Childcare for working women became a political issue and a government decision soon followed offering childcare for all children, signalling the modern history of Swedish ECE (Martin Korpi, 2016). In the Commission on Nursery Provision (SOU, 1972: 26), directives were drafted to guide municipal childcare planning. From the First Preschool Act 1975, childcare provision became a mandatory municipal task, where municipalities were obliged to take responsibility for expansion. Alongside a government terminology shift from childcare to preschool, organisational directives were provided for the preschools and their everyday curricula. Such directives aligned with the then contemporary child development research (e.g. Piaget, 1953, 1969) and Freire's work (1970, 2021) that promoted progressive pedagogies to address inequities. The Commission argued that a preschool curriculum could create equal conditions for all children's learning and development.

The Commission on Nursery Provision (SOU, 1972: 26) introduced a nature orientation to strengthen the development of scientific knowledge as a fundamental preschool activity. The Commission argued for the development of children's understanding of scientific phenomena and knowledge through everyday encounters with flora and fauna. In addition, environmental protection was included as important content for teachers to share with children. This reflected increased societal awareness about environmental issues evident in the first world environment conference, the UN Stockholm conference in 1972 (United Nations Conference on the Human Environment, 1972). The preschool nature and EE curriculum content were further clarified in a steering document, the *Pedagogical Program for Preschool* (National Board of Health & Welfare, 1987). It stated that the preschool should work with content knowledge about plants and animals, climate and environment, natural phenomena, nature conservation, and mathematical and technical understandings. Other proscribed content included the human body, health and well-being, people's similarities and differences, gender identity and emotions. Overall, it recommended children be offered opportunities to develop understandings of society's common environmental problems through activities in and about nature. In the *Pedagogical Program for Preschool,* waste sorting (paper recycling) was described for the first time as curriculum content and suggestions to manage any anxiety children may feel about environmental problems were included. From 1987, developing environmental awareness was described as crucial and involved activities such as caring for plants or growing food. A positive attitude towards environmental issues was advocated, so that preschool children—in everyday contexts and situations—learnt it was possible to do something about environmental problems.

Contemporary policy shifts: The next iteration of the *Swedish National Preschool Curriculum* (SOU, 1997:157) was strongly influenced by *Agenda 21* (1993). Agenda 21 was a non-binding action plan proposed by the United Nations for global, national, and local uptake by organisations across every area of human environmental impact. Preschools were to have an important educative role in efforts towards globally sustainable futures, thus the curriculum described environmental practices and content from local and global perspectives. The local perspective focused on contact with nature and utilising play as a learning approach for developing children's understandings and knowledge about nature and the environment. The intent was an emphasis on children's own discoveries and exploration in nature as essential to human-nature connections. The children were encouraged to develop a positive belief in the future for ultimately contributing towards an ecologically sustainable society. From a global perspective, lifestyle issues, global justice, and problematising the sustainability of the Earth's resources were specified as important. Although the first *Swedish National Preschool Curriculum* was introduced in 1998 (Swedish National Agency for Education, 1998), some parts have not yet been realized. The global perspective was later reduced, in favour of an ecological approach to everyday life.

> The preschool shall place great emphasis on environmental and nature conservation issues. An ecological approach and a positive belief in the future should characterize the preschool's activities. The preschool should contribute to the children acquiring a careful approach to nature and the environment and understanding their participation in nature's cycle... to help children understand how everyday life and work can be designed so that it contributes to a better environment both in the present and in the future. (Swedish National Agency for Education, 1998, p. 7)

In addition, the concept EfS or learning for SD was not employed in the curriculum document. A decade later, this concept was included in the revised Swedish National Preschool Curriculum (Swedish National Agency for Education, 2018, 2019) and it specifically highlighted:

> Education should be characterised by a positive belief in the future. Education should give children the opportunity to acquire an ecological and caring approach to their surrounding environment and to nature and society. Children should also be given the opportunity to develop knowledge about how the different choices that people make can contribute to sustainable development – not only economic, but also social and environmental. (2019, p. 10)

There is also a long tradition of embedding fundamental values:

> respect for the inviolability of human life, individual freedom and integrity, the equal value of all people, equality between women and men, girls and boys, and solidarity between people (2019, p. 5)

According to the curriculum, the work on fundamental values in preschools is possible as children develop their abilities to be considerate and demonstrate respect and care for others and the environment. Ethical dilemmas that arise in everyday preschool settings can involve taking a stand and ethics discussions offer multiple ways of understanding humans and their environments.

Influence of other actors with an interest in nature, environment and sustainability: Alongside the policies discussed above, other actors have exerted considerable influence, enhancing education towards sustainability in preschools. For example, *Skogsmulleskolan*, the pedagogical program for children's outdoor play and learning began in the late 1950s. The activities were based on children's interests and being in nature to develop knowledge about nature and an environmental consciousness, plus a healthy lifestyle. From the 1970s onwards, preschool staff were offered training to lead forest activities. Later, in the 1980s the worldwide *Come Rain or Shine* movement instigated by the NGO (The Swedish Outdoor Promotion Association [Friluftsfrämjandets]) was established. There are around two hundred Swedish Come Rain or Shine preschools; and, their ethos cherishes children's play, discovery, and curiosity outdoors (Gustavsson & Söderberg, 2021).

Another influential initiative in the late 1980s was the National Board of Health and Welfare funding opportunity for municipalities to work with environmental issues and education in preschools. This initiative resulted in several books on EE in the preschool and outdoor education (see Holmström et al., 1992).

Furthermore, after the Rio Agenda 21 conference, offices were established in each Swedish municipality to promote sustainability issues. Forsberg (2002) analysed this development and commented on the surprise expressed by an Agenda 21 coordinator in evaluating this work. 'Where at first I thought things would happen—The municipal environmental office and waste sorting—nothing happened. On the other hand, we didn't think anything special about childcare, but there everything took off at a breathtaking speed' (Forsberg, 2002, p. 158 [translated from Swedish]).

Other interventions were the establishment of national awards to stimulate preschools and schools to work with sustainability and two major certifications available to Swedish preschools. Firstly, the 'Green Flag', instigated with support from the Keep Sweden Tidy Foundation in 1996, as part of the international Foundation for Environmental Education (FEE) Eco-schools. The second award, 'School for Sustainable Development', was first known as the 'Environmental School' award and began in 1998 as a government initiative administered by the Swedish National Agency for Education. Notably, this award takes a broader perspective on sustainability (social, economic, and ecological), while the Green Flag focuses primarily on environmental work and EE.

Over the years, various forms of collaboration about sustainability issues have emerged with preschools. For example, the Keep Sweden Tidy Foundation arranges annual environmental cleaning with preschools and has conducted training for preschool staff and developed educational materials (Keep Sweden Tidy Foundation, 2023). More recently, other actors such as the World Wide Fund for Nature (WWF) have developed preschool training and educational materials around a sustainability theme. Lastly, research has been another influential factor. Swedish research about EE and sustainability developed somewhat late considering the ongoing curricular policy interest. Many Swedish preschools currently work with sustainability issues

(Ärlemalm-Hagsér, 2013; Borg, 2017; Larsson & Pramling Samuelsson, 2019), but there is still a distinct lack of field studies (Ärlemalm-Hagsér & Hedefalk, 2018; Ärlemalm-Hagsér & Sundberg, 2016). Such studies have recently grown in number (e.g. Ärlemalm-Hagsér & Pramling Samuelsson, 2021; Borg & Samuelsson, 2022; Caiman et al., 2022) and several books (Engdahl & Ärlemalm-Hagsér, 2020; Pramling Samuelsson et al., 2021) have been published to support preschool staff in developing their pedagogical practices aligned with the national curriculum (Swedish National Agency for Education, 2018, 2019). This both extends and consolidates the relatively long history of Swedish preschools working with sustainability issues.

In summary, this modern history of Swedish preschools, alongside the high standards set by the Swedish Government (SOU, 2019:13) and the National Preschool Curriculum (Swedish National Agency for Education, 2018, 2019) demonstrates clear sustainability goals. Yet, in contemporary studies (Ärlemalm-Hagsér, 2022; Engdahl et al., 2021), Swedish preschool teachers are still struggling to understand the task and how to handle the complexity and wickedness of sustainability problems.

2.3.3 International Landscapes: History, Research, and Policy

The Australian and Swedish ECEfS histories outlined above are very much enmeshed with international historical landscapes where key individuals and organisations from both countries have significantly contributed to international uptake. We argue this has been very much a two-pronged approach raising awareness of the potential for ECE to have a leading role in the EfS/ESD field and alerting the ECE field to the importance of embracing EfS/ESD for children's sustainable futures.

Swedish Professor Pramling Samuelsson was an early leader on the international stage and effectively adopted this two-pronged approach. As Chair of Early Childhood Education and Sustainable Development at UNESCO since 2008, and World president of OMEP, in 2007 she instigated the first UNESCO ESD workshop of international delegates focused on ECE (see Chap. 5); notably, co-authors Davis and Ärlemalm-Hagsér were participants. An outcome was the first international publication addressing ECEfS (Pramling Samuelsson & Kaga, 2008) which stated, 'it is not necessary to invent 'new' pedagogies to 'do' education for sustainability in the early years—one can build on its pedagogical traditions to do so' (p. 13). From this defining point during the UNDESD 2005–2014, the potential of ECE as integral to EfS/ESD was clearly recognised on an international stage and supported by successive international research studies, conferences, publications, organisations, and policy shifts. Illustrative of these expanding international horizons is the third edition of *Young Children and the Environment* (Davis & Elliott, 2024) which included four chapters by international co-authors beyond Australia.

2.3.4 *From a Research Hole to Building an International Research Base*

In the initial decades of the 1980s, 1990s, and early 2000s, there was virtually no research to inform practice or policy. In 2009, as previously noted, Davis alerted researchers to this 'hole'. This led directly to the instigation of the Transnational Dialogues in ECEfS Research (TND), for addressing this issue. In addition, illustrative of Davis's concerns at the time the *International Handbook of Research on Environmental Education* (Stevenson et al., 2012), a 592-page tome with 51 chapters and over 100 authors, did not include one single chapter on ECEfS/ESD. Many questions can be posed here, about oversight, unwitting biases, the alternative lens of ECE, or simply a nascent ECEfS/ESD research field not yet visible to other researchers. Regardless, these questions continue to capture our challenges over several decades of advocacy. Only in last the decade has a strong research base been established offering a rich canvas of theoretical, methodological, and pedagogical insights into ECEfS/ESD (see Chaps. 3 and 4). The TND and European Early Childhood Education Research Association (EECERA) Sustainability SIG research groups have been instrumental here.

Transnational Dialogues (TND): Since 2010, the TND, a research group established by Professor Julie Davis (Australia) and Professor Eva Johansson (Norway) has been driving significant international networking. Initially, the group involved a small number of invited participants known to the instigators from Australia, Norway, New Zealand, and Sweden. Each TND gathering offers varied research foci, invited speakers, experienced and emerging researchers, and generative discussion incorporating local and global perspectives. As it is self-funded, every effort has been made over the years to align TND gatherings with a major international conference, most notably the biennial World Environmental Education Congress (WEEC) or the annual EECERA conference. Both now have a significant ECEfS/ESD presence, for example, WEEC has incorporated an ECE theme since 2011. These large international conferences became gathering places for new participants, especially as TND researchers presented symposia and conference papers in addition to the TND meetings themselves. It is noted that the current convenors are from Australia, Norway (original counties of origin) as well as Sweden and Turkey. With COVID, the network took the opportunity to meet online; this radically opened the TND to larger numbers of researchers from many more countries, as shown in Table 2.1 from TND 5 onwards. The promotion of research collaborations and publications—with explicit emphasis on publishing research papers and books—has been a hallmark of the TND. It is TND members who have taken the lead in publishing many of the outputs illustrated in Tables 4.2, 4.3 and 4.6.

A TND feature has been the online collation of a Biobook comprising biographical and research interest details for each TND participant to facilitate connections. An analysis of the Biobook text revealed a shared participant focus on urgency, equity, and agency (Dean & Elliott, 2022). The current TND mailing list includes over 150 participants representing countries from Asia to Europe, Scandinavia, the

Table 2.1 Overview of TND gatherings 2010–2023

TND dates	Location	Participant numbers
TND 1 August 2010	University of Stavanger, Stavanger, Norway (after 26th OMEP World Congress)	15
TND 2 July 2011	Queensland University of Technology, Brisbane, Australia (before 6th WEEC)	25
TND 3 July 2015	University of Stavanger, Stavanger, Norway (before 8th WEEC)	25
TND 4 September 2017	University of Victoria, Victoria, Canada (before 9th WEEC)	22
TND 5.1 & 5.2 September 2020	Online	61
TND 6.1 September 2021 TND 6.2 March 2022	Online Prague Czech Republic (before 11th WEEC)	35 7
TND 7.1 August 2022 TND 7.2 October 2022	Glasgow, UK, (before 30th EECERA, combined with EECERA Sustainability SIG) Online	24 30
TND 8 August 2023	Mälardalen University, Västerås, Sweden (before 31st EECERA)	27

Americas, Africa, and Australasia. What once began as primarily a Scandinavian and Australasian initiative is now global.

Emery et al. (2017) celebrated the purpose, rationale, and significance of this evolving TND network. The co-authors described TND as both a network model and dialogic process for fostering collaborative research initiatives and building researcher capacity. Through personal connections and gathering like-minded others along the way, TND has become a rich multinational, multidisciplinary space for sharing ECEfS perspectives and generating research outcomes. Published collaborative TND outcomes include two edited research volumes, Davis and Elliott (2014) *Research in early childhood education for sustainability: International perspectives and provocations* and Elliott et al. (2020) *Researching early childhood education for sustainability: Challenging assumptions and orthodoxies*. In addition, through TND connections a raft of research publications—papers, special issues, and books—can now be attributed to those in some way connected with TND (Ärlemalm-Hagsér & Elliott, 2017; Carr et al., 2021; Davis & Elliott, 2024; Harwood et al., 2020; O'Gorman, 2017) (see Chaps. 3 and 4).

EECERA SIG: The above-mentioned EECERA Sustainability SIG was established in 2019 by founding co-convenors Ärlemalm-Hagsér, Furu, Alici, and Elliott. The underpinning argument is that it was timely for all mainstream ECE bodies to recognise the urgency of global sustainability and demonstrate leadership (https://www.eecera.org/sig/sustainability-in-early-childhood-education/). We strongly argue there is an opportunity for every ECE organisation, whether a management group or professional association to advocate for ECEfS, particularly through sustainability policies and position statements. An EECERA SIG sustainability position statement was ratified and launched at the 2023 conference to promote ECEfS within and beyond EECERA. The SIG group has over 90 members and meets annually at EECERA conferences where ECEfS/ESD symposia are regularly presented. In addition, the SIG hosts regular online forums, develops publications (Evans et al., 2022), and shares a Facebook site for postings (https://www.facebook.com/groups/724907741352504/).

Overall, the ECEfS international research base is steadily building and the 'hole' is filling. We acknowledge Chap. 5 respondents as key to this endeavour. Increasingly, mainstream journals have published special issues on ECEfS/ESD (see Chap. 4). In 2012, the North American Association for Environmental Education established a dedicated *International Journal of Early Childhood Environmental Education* and in 2013, instigated the *Natural Start Alliance*, another forum for promoting ECE research and practice. Literature reviews (Ardoin & Bowers, 2020; Hedefalk et al., 2014; Somerville & Williams, 2015), since Davis's initial 2009 clarion call, attest to the increasing number of ECEfS/ESD research studies reported and the breadth of theoretical and methodological framing now evident.

2.3.5 International Policies and ECEfS

Nevertheless, acknowledgement of the critical role of ECE in EfS/ESD within international policies has often lagged behind the shifts described in the historical landscapes of Australian and Swedish ECEfS/ESD around curriculum, networks, and publications dating back to the 1980s and 90s.

United Nations and UNESCO: The first acknowledgement, the UNESCO (2008) *Gothenburg Recommendations on Education for Sustainable Development*, unequivocally stated, 'It is imperative that Early Childhood Education (ECE) is recognized as the starting point for lifelong learning within education for sustainability' (p. 26). This statement is now reflected upon as foundational to later policies, initiatives, and reports such as the *UNESCO Decade of Education for Sustainable Development 2005–2014* (UNDESD) (2005). It is significant that when the UNDESD began in 2005 ECE was largely absent, but by decade end the final UNESCO report stated: 'ECCE [Early Childhood Care and Education] has a clear role in preparing present and future citizens and in aiding societies to make the necessary transitions to sustainability' (2014, p. 70). Co-authors Davis and Elliott reflect that in their sustainability work with Australian practitioners, the UNDESD (UNESCO, 2005) was largely unknown throughout the entire decade. Yet in 2012, in response to requests from the global field, UNESCO published a set of case studies about EC ESD good practice (UNESCO, 2012). Unfortunately, the UNDESD follow-up, the *UNESCO Global Action Programme on Education for Sustainable Development* (2015) showcased many global initiatives involving schools and communities, but none specifically drawn from the ECE sector demonstrating ECE was not yet consistently at the forefront.

However, Agenda 30 and the Sustainable Development Goals (UN, 2015) appear to have gained more traction in the ECE field. As a global initiative the ECE focus on SDG4: Quality Education is most often equity, access, and participation and while there are critiques of the SDGs, there is much scope for interpreting all SDGs in ECE settings. We note SDG 4 Target 4.7 highlights the demand that all learners (including early childhood learners) are exposed to ESD. To this end, publications by Carr et al. (2021) and Sonter et al. (2024) offer both research and practice insights about the SDG's in ECE settings.

Creating international momentum for change: In addition, to the research and policy foci above, OMEP has been a key player in creating global networking opportunities for change (see Chap. 5). OMEP as an international advocacy body with membership across 70 countries has implemented projects and awards and embedded EfS/ESD as a conference priority over more than a decade. The OMEP ESD rating scale (Siraj-Blatchford et al., 2016) offered a starting point for assessing EfS/ESD implementation in ECE contexts and has been revised and adopted in various countries (Nikiforidou et al., 2019; OMEP, 2019). From these pivotal beginnings, OMEP has offered and continues to offer international organisational leadership.

The most recent prioritising of EC EfS/ESD is evident in the reports of the European Union Commission (2022), The International Commission on the Futures of Education (2021), and Conference of Parties (CoP 26) (2021). The 2022 EU Commission Proposal for a Council Recommendation on Learning for Environmental Sustainability urges member states to increase their efforts 'To develop its full potential, learning and teaching for environmental sustainability needs to happen not only in schools and higher education but in all parts of the system (formal, non-formal, informal) and at all levels (from early childhood to adulthood into older age)' (EU Commission, 2022, p. 2). Essentially this reiterates the stance of the earlier Gothenburg Recommendations (UNESCO, 2008); there is no debate that ECE plays a key role as the starting point for ongoing EfS/ESD. The International Commission on the Futures of Education (2021) while not explicitly mentioning ECE recognises 'a pressing need to reimagine why, how, what, where, and when we learn – meaning that education is not yet fulfilling its promise to help us shape peaceful, just, and sustainable futures' (p. 6). We argue that ECE offers a foundation for the innovative and regenerative approaches to education called for to build more sustainable and just societies.

The 2021 CoP 26 event in Glasgow included for the first time a 'Joint Event of Education and Environment Ministers Summit - Together for Tomorrow: Education and Climate Action' chaired by Professor Daniella Tilbury. Central to this meeting was the drafting of national pledges and we noted repeated reference to EfS/ESD needing to start in ECE (see Chap. 6). Lastly, recent interest in such foci includes the Asia–Pacific Regional Network for Early Childhood (ARNEC, https://arnec.net/) and newcomer This Is Planet Ed (USA Aspen Institute advocacy group, https://www.thisisplaneted.org/). Key international organisations, reports, and events offer leverage points at national and local levels for urgent action, whether it be policy shifts, funded initiatives, educator training, or research. Momentum around ECEfS/ESD is being created at last.

2.4 Reflecting on Strengths and Tensions Over Time

As co-authors and participants in this shared history, we now reflect briefly on the strengths and tensions we have experienced. Most ardently, we argue for the strength of the networks at multiple levels, locally, nationally, and internationally, in driving the ECEfS/ESD agenda forward. The networks generated self-sustaining Communities of Practice where all perspectives were welcomed and richness in diversity celebrated. As the international network representation has widened across majority/minority worlds, we have all critically reflected on the culturally varied meanings of sustainability. First hand, we have experienced over many years the collegiality and academic generosity of these networks, a lasting memory for us all. As long-term ECEfS/ESD advocates recognising the pressing concerns of climate change (IPCC, 2023) and children's rights to sustainable futures, we regularly experience tension around the slow uptake of ECEfS/ESD in the field and lagging policy change. We are accustomed to celebrating small milestones, as we continue to strongly advocate for transformative global change around ECEfS/ESD.

Appendix: A Chronology of Key International ECEfS Publications, Studies and Events 1987–2023

Decade	Event or publication	Origin
1970s	United Nations (UN). (1972). United Nations conference on the Human Environment, 5–16 June 1972, Stockholm	International
1980s	World Commission on Environment and Development (WCED) (1987). *Brundtland report: Our common future.* Oxford University Press	International
1990s	Elliott, S., & Emmett, S. (1991). *Snails live in houses too, environmental education for the early years.* Martin Educational. (2nd ed., 1997)	Australia
	Environmental Education in Early Childhood Victoria. (EEEC Vic Inc) established 1992, ongoing, www.eeec.com.au	Australia
	Wilson, R. (1993). *Fostering a sense of wonder during the early childhood years.* Greyden Press	USA
	Queensland Early Childhood Environmental Education Network established 1995, now Queensland Early Childhood Sustainability Network (QECSN)	Australia

(continued)

(continued)

Decade	Event or publication	Origin
2000s	New South Wales Environment Protection Authority (NSW EPA) (2003). *Patches of green: A review of early childhood environmental education*. NSW EPA	Australia
	Davis, J., & Elliott, S. (2003). *Early childhood environmental education: Making it mainstream*. Early Childhood Australia	Australia
	Davis, J. (2005). Education for sustainability in the early years: Creating cultural change in a childcare setting. *Australian Journal of Environmental Education, 21*, 47–55	Australia
	Davis, J., Gibson, M., Pratt, R., Eglington, A., & Rowntree, N. (2005). Creating a culture of sustainability: From project to integrated education for sustainability at Campus Kindergarten. In W.L. Filho (Ed.), *The International Handbook of Sustainability Research* (pp. 563–594). UNESCO	Australia
	UNESCO (2005). *Decade of education for sustainable development 2005–2014 Draft International Implementation Scheme*. UNESCO	International
	Tilbury, D., Coleman, V., & Garlick, D. (2005). *A national review of environmental education and its contribution to sustainability in Australia: School education*. Australian Government Department of the Environment & Heritage and Australian Research Institute in Education for Sustainability. (Sections on ECEfS by Davis and Elliott)	Australia
	Inaugural Australasian Early Childhood Education for Sustainability Conference 2006, Christchurch, New Zealand	Australia/ New Zealand
	Pramling Samuelsson, I. (2006). Sustainable development – what does that mean in preschool? In A. Pipere (Ed.), *Education & sustainable development. First steps toward changes*, Vol 1, (pp. 182–187). Daugavpils University Publishing House "Saule"	Sweden
	UNESCO (2008). *The Gothenburg recommendations on education for sustainable development*. UNESCO	International
	Pramling Samuelsson, I., & Kaga, Y (2008). *Early childhood and its contribution to a sustainable society*. UNESCO	International
	Davis, J. (2009). Revealing the research 'hole' of early childhood education for sustainability: A preliminary survey of the literature. *Environmental Education Research, 15*(2), 227–241	International
	Elliott, S., & Davis, J. (2009). Exploring the resistance: An Australian perspective on educating for sustainability in early childhood. *International Journal of Early Childhood, 41*(2), 65–77. (Special themed edition on sustainability)	International
	Pramling Samuelsson, I. (2011). Why we should begin early with ESD: The role of early childhood education. *International Journal of Early Childhood, 43*(2), 103–118. http://www.springer.com/education+%26+language/journal/13158	Sweden
2010s	Davis, J. (Ed.) (2010). *Young children and the environment: Early education for sustainability*. Cambridge Press. Translated into Korean (2014) and Chinese (2017)	International

(continued)

(continued)

Decade	Event or publication	Origin
	Siraj-Blatchford, J., Smith, K. C., & Pramling Samuelsson, I. (2010). *Education for sustainable development in the early years*. OMEP	UK Sweden
	Ritchie, J. (2010). Fostering communities: Ecological sustainability within early childhood education. *Early Education 47* Autumn/Winter, 10–14	New Zealand
	Elliott, S. (2010). Essential not optional: Education for sustainability in early childhood centres. *Childcare Information Exchange*. March/April, 34–37	Australia/USA
	First Transnational Dialogues in Early Childhood Education for Sustainability Research (TND) in 2010, with founding co-conveners Eva Johansson and Julie Davis and hosted by University of Stavanger, Norway	Australia Norway
	Siraj-Blatchford, J. et.al. (2010). Taking children seriously – How the EU can invest in early childhood education for a sustainable future. *European Panel of Sustainable Development, EPSD. Report No 4, 2010–12-17*. GMV, Centre for Environment and Sustainability: Chalmers U. of Technology. www.ecesustainability.org	International
	North American Association for Environmental Education (2010). *Early childhood environmental education programs: Guidelines for excellence*. NAAEE	USA
	UNESCO DESD *Reorientating Teacher Education Forum*, 2010 including Early Childhood Education	International
	World Environmental Education Congress, 2011, Brisbane, included an early childhood strand, plus 2nd Transnational Dialogues in Early Childhood Education for Sustainability Research	International
	Edwards, S., & Cutter-Mackenzie, A. (2011). Environmentalising early childhood education curriculum through pedagogies of play. *Australasian Journal of Early Childhood [E], 36*(1), 51–59	Australia
	North American Association for Environmental Education (NAAEE) Established in 2012 the *International Early Childhood Environmental Education Journal* www.naaee.org.us	International
	United Nations Educational, Scientific and Cultural Organisation (UNESCO) (2012). *Education for sustainable development: Good practices in early childhood*. UNESCO. https://unesdoc.unesco.org/ark:/48223/pf0000217413	International
	Elliott, S., Edwards, S., Davis, J., & Cutter-McKenzie, A. (Eds.) (2013). *Best of sustainability: Research, theory and practice*. Early Childhood Australia	International
	Kelly, J., & White, E. J. (2013). *The Ngahere Project: Teaching and learning possibilities in nature settings*. Wilf Malcolm Institute of Educational Research, University of Waikato. http://www.waikato.ac.nz/data/assets/pdf_file/0007/146176/Ngahere-project_3-2013-03-14.pdf	New Zealand

(continued)

(continued)

Decade	Event or publication	Origin
	Davis, J., & Elliott. S. (Eds.) (2014). *Research in early childhood education for sustainability: International perspectives and provocations*. Routledge. First TND publication and republished in Korean (2014)	International
	Elliott, S., & Young, T. (2015). Nature by default in early childhood education for sustainability. *Australian Journal of Environmental Education, 32*(1), 57–64. https://doi.org/10.1017/aee.2015.44	Australia
	Williams, C., & Somerville, M. (2015). Sustainability education in early childhood: An updated review of research in the field. *Contemporary Issues in Early Childhood, 16*(2), 102–117. https://doi.org/10.1177/1463949115585658	International
	Hedefalk, M., Almqvist, J., & Östman, L. (2015). Education for sustainable development in early childhood education: a review of the research literature. *Environmental Education Research, 21*(7), 975–990. https://doi.org/10.1080/13504622.2014.971716	International
	United Nations (UN). (2015). *Transforming our world: The 2030 agenda for sustainable development*. http://www.un.org/ga/search/view_doc.asp?symbol=A/RES/70/1&Lang=E	International
	Currie, J., & Deschenês, O. (2016). Children and climate change: Introducing the issue. *The Future of Children, 26*(1), 3–9. https://www.jstor.org/stable/43755227	International
	Elliott, S., McCrea, N., Newsome, L., & Gaul, J. (2016). *Examining environmental education in NSW early childhood education services: A literature review with findings from the field*. NSW OEH Environmental Trust. http://www.environment.nsw.gov.au/grants/dissemination.htm	Australia
	Siraj-Blatchford, J., Mogharreban, C., & Park, E. (2016). *International research on education for sustainable development in early childhood*. Springer	International
	Ärlemalm-Hagsér, E., & Elliott, S. (2017). Special Issue: Contemporary Research on Early Childhood Education for Sustainability. *International Journal of Early Childhood, 49*(3), 267–272	International
	Cutter-Mackenzie-Knowles, A., Malone, K., & Barratt Hacking, E. (Eds.). (2018). *International research handbook on childhoodnature: Assemblages of childhood and nature research*. Springer	International
	Boyd, D., Hirst, N., & Siraj-Blatchford, J. (2018). *Understanding sustainability in early childhood education: Case studies and approaches from across the UK*. David Fulton	UK
	Huggins, V., & Evans, D. (Eds.) (2018). *Early childhood education and care for sustainability: International perspectives*. Routledge	International
	OMEP (2019). *ESD Rating scale* (2nd ed.) https://omepworld.org/education-3/	International

(continued)

(continued)

Decade	Event or publication	Origin
	Inaugural EECERA Sustainability Special Interest Group meeting, Thessaloniki, Greece Conference, August 2019	International
	Smidt, S. (2019). *Early childhood education and care for a shared sustainable world.* Routledge	UK
2020s	Elliott, S., Ärlemalm-Hagsér, E., & Davis, J. (Eds.) (2020). *Researching early childhood education for sustainability: Challenging assumptions and orthodoxies* (Vol. 2). Routledge. Second TND publication	International
	Ardoin, N., & Bowers, A. (2020). Early childhood environmental education: A systematic review of the research literature. *Education Research Review, 31*, 1–16. https://doi.org/10.1016/j.edurev.2020.100353	USA
	Clark, H. et al., (2020). A future for the world's children? WHO–UNICEF–Lancet Commission. *The Lancet Commissions, 395*(10,224) 605–658. https://doi.org/10.1016/S0140-6736(19)32540-1	International
	Güler Yıldız, T., Öztürk, N., İlhan İyi, T., Aşkar, N., Banko Bal, Ç., Karabekmez, S., & Höl, Ş. (2021). Education for sustainability in early childhood education: A systematic review. *Environmental Education Research, 27*(6), 796–820. https://doi.org/10.1080/13504622.2021.1896680	International
	Carr, V., Elliott, S., & Ärlemalm-Hagsér, E. (2021). Editorial Special Issue: *International Journal of Early Childhood Environmental Education - Sustainable Development Goals and Early Childhood Environmental Education: Awareness, Understanding, and Transformation, 9*(1), 5–8. https://naturalstart.org/research/ijecee/volume-9-number-1	International
	Engdahl, I,. Pramling Samuelsson, I., & Ärlemalm-Hagsér, E. (2021). Swedish Teachers in the Process of Implementing Education for Sustainability in Early Childhood Education. *New Ideas in Child and Educational Psychology, 1*(1), 3–23	Sweden
	International Commission on the Futures of Education (2021). *Futures of education: Progress update.* UNESCO. https://unesdoc.unesco.org/ark:/48223/pf0000375746/	International
	United Nations Educational, Scientific and Cultural Organisation (UNESCO) (2021). *Re-imagining our futures together: A new social contract for education.* UNESCO	International
	Evans, N. Harwood, D., Furu, A.-C., & Alici, Ş. (2022). Challenges in reorienting early childhood education for sustainability in initial teacher education: Transforming, creating and expanding. *Special Issue: The Australian Journal of Teacher Education, 47*(5), https://doi.org/10.14221/ajte.2022v47n5.1	International

(continued)

(continued)

Decade	Event or publication	Origin
	Engdahl, I., & Furu, A-C. (2022). Early Childhood Education for Sustainability. A Vibrant Arena in the Complex Transformation of Society Towards Sustainability. *International Journal of Early Childhood, 54*(1), 1–12. https://doi.org/10.1007/s13158-022-003 23-0	International
	Engdahl, I., Pramling Samuelsson, I., & Ärlemalm-Hagsér, E. (2023). Systematic child talks in early childhood education—A method for sustainability. *Children, 10*, 661. https://doi.org/10.3390/children10040661	Sweden
	Rooney, T., & Blaise, M. (2023). *Rethinking environmental education in a climate change era: Weather learning in early childhood*. Routledge	Australia
	Elliott, S., & Davis, J. (2023). *Early Childhood Australia Essentials: Sustainability in early childhood*. Early Childhood Australia	Australia
	EECERA Sustainability Special Interest Group Launch of Position Statement, Estoril, Portugal Conference, August 2023	International
	Eighth Transnational Dialogues in Early Childhood Education for Sustainability Research (TND) in 2023, hosted by Mälardalen University, Sweden	International
	Davis, J., & Elliott, S. (2024). *Young children and the environment: Early education for sustainability* (3rd ed.). Cambridge University Press	International

References

Australian Curriculum, Assessment and Reporting Authority (ACARA). (2020). *Australian National Curriculum Sustainability Cross Curriculum Priority*. ACARA. https://www.australiancurriculum.edu.au/f-10-curriculum/cross-curriculum-priorities/sustainability/

Australian Children's Education and Care Quality Authority (ACECQA). (2011 [2013]) *Guide to the National Standard.* ACECQA.

Australian Children's Education and Care Quality Authority (ACECQA). (2018). *Guide to the National Standard.* ACECQA.

AGDE. (2022). *Belonging, being and becoming: Early years learning framework for Australia (V2.0).* Australian Government Department of Education for the Ministerial Council.

Australian Productivity Commission. (2014). *Childcare and early childhood learning. Inquiry Report No. 73.* Australian Productivity Commission.

Ardoin, N., & Bowers, A. (2020). Early childhood environmental education: A systematic review of the research literature. *Education Research Review, 31*, 1–16. https://doi.org/10.1016/j.edurev.2020.100353

Ärlemalm-Hagsér, E. (2013). *An interest in the best for the world'? Education for sustainability in the Swedish preschool* [Doctoral thesis, Gothenburg Studies in Educational Sciences 335]. Gothenburg: Acta Universitatis Gothoburgensis.

Ärlemalm-Hagsér, E., & Davis, J. (2014). Examining the rhetoric: A comparison of how sustainability and young children's participation and agency are framed in Australian and Swedish early childhood education curricula. *Contemporary Issues in Early Childhood, 15*(3), 231–244. https://doi.org/10.2304/ciec.2014.15.3.231

Ärlemalm-Hagsér, E., & Elliott, S. (2017). Special Issue: Contemporary Research on early childhood education for sustainability. *International Journal of Early Childhood, 49*(3), 267–272.

Ärlemalm-Hagsér, E., & Hedefalk, M. (2018). Editorial: Förskolan och utbildning för hållbarhet–Nordisk forskning inom fältet förskola och hållbarhet [Editorial Special Issue: Contemporary Nordic Research on Early Childhood Education for Sustainability. *Education & Democracy, 27*(2), 7–14.

Ärlemalm-Hagsér, E., & Sundberg, B. (2016). Nature experiences and recycling—A quantitative study on education for sustainable development in Swedish preschools. *NorDina, 12*(2), 140–156.

Ärlemalm-Hagsér, E., & Pramling Samuelsson, I. (2021). "Business as usual"? Or transformative and transactive teaching leading towards the Agenda 2030 goals in Swedish Early Childhood Education. *International Journal of Early Childhood Environmental Education, 9*(1), 94–111.

Ärlemalm–Hagsér, E. (2022). Förskolans utbildning i antropocen [Preschool education in the Anthropocene]. *Pedagogisk forskning i Sverige. 27*(3), 96–117.

Borg, F. (2017). *Caring for people and the planet: preschool children's knowledge and practices of sustainability* [Thesis, Umeå University, Sweden].

Borg, F., & Pramling Samuelsson, I. (2022). Preschool children's agency in education for sustainability: The case of Sweden. *European Early Childhood Education Research Journal.* https://doi.org/10.1080/1350293X.2022.2026439

Caiman, C., Hedefalk, M., & Ottander, C. (2022). Pre-school teaching for creative processes in education for sustainable development—Invisible animal traces, purple hands, and an elk container. *Environmental Education Research, 28*(3), 457–475. https://doi.org/10.1080/13504622.2021.2012130

Carr, V., Elliott, S., & Ärlemalm-Hagsér, E. (2021). *Editorial Special Issue: International Journal of Early Childhood Environmental Education—Sustainable Development Goals and Early Childhood Environmental Education: Awareness, Understanding, and Transformation, 9*(1), 5–8. https://naturalstart.org/research/ijecee/volume-9-number-1

Commonwealth of Australia Department of Education, Employment & Workplace Relations (DEEWR). (2009). *Belonging, being and becoming: Early years learning framework for Australia.* Commonwealth Department of Education, Employment & Workplace Relations.

Commonwealth of Australia Department of Environment & Heritage (DEH). (1999). *Today shapes tomorrow: Environmental education for a sustainable future.* Commonwealth Department of Environment & Heritage.

Conference of Parties (CoP26). (2021). *The joint event of education and environment ministers Summit—together for tomorrow: Education and climate action.* https://lnkd.in/d-8drcy7

Council of Australian Governments (COAG). (2008). *National Quality Framework. A national quality framework for early childhood education and care, a discussion paper.* COAG Productivity agenda working group: Education, skills, training and early childhood development. https://www.acecqa.gov.au/sites/default/files/2020-12/ANationalQualityFrameworkForEarlyChildhoodEducation.pdf

Davis, J. (1998). Young children, environmental education, and the future. *Early Childhood Education Journal, 26*, 117–123. https://doi.org/10.1023/A:1022911631454

Davis, J. (2005). Education for sustainability in the early years: Creating culture change in a child care setting. *Australian Journal of Environmental Education, 21*, 7–55.

Davis, J. (2009). Revealing the research 'hole' of early childhood education for sustainability: A preliminary survey of the literature. *Environmental Education Research, 15*(2), 227–241. https://doi.org/10.1080/13504620802710607

Davis, J. (Ed.). (2010). *Young children and the environment: Early education for sustainability.* Cambridge Press.

Davis, J. (Ed.). (2015). *Young children and the environment: Early education for sustainability* (2nd ed.). Cambridge University Press.

Davis, J. M., & Davis, J. E. (2020). Early childhood teacher education and education for sustainability: A review of the literature and mapping of courses. In S. Elliott, E. Ärlemalm-Hagsér, & J. Davis (Eds.), *Researching early childhood education for sustainability: Challenging assumptions and orthodoxies* (Vol. 2) (pp. 111–124). Routledge.

Davis, J., & Elliott, S. (2003). *Early childhood environmental education: Making it mainstream.* Early Childhood Australia.

Davis, J., & S. Elliott (Eds.). (2014). *Research in early childhood education for sustainability: International perspectives and provocations.* Routledge.

Davis, J., & Elliott, S. (2024). *Young children and the environment: Early education for sustainability* (3rd ed.). Cambridge University Press.

Davis, J., Miller, M., Boyd, W., & Gibson, M. (2008). *The impact and potential of water education in early childhood care and education settings.* Brisbane, Queensland: Rous Water Board, Queensland University of Technology and Institute of Sustainable Resources.

Dean, S., & Elliott, S. (2022). Urgency, equity, and agency: An assemblage of global concerns and interests in education for sustainability. *International Journal of Early Childhood Environmental Education, 9*(2), 56–68.

Dighe, J. (1993). Children and the Earth. *Young Children, 48*(3), 58–63.

Early Childhood Australia (ECA). (2016). *Code of Ethics.* https://www.earlychildhoodaustralia.org.au/our-publications/eca-code-ethics/

Early Childhood Australia. (2021). *Provocations for our Earth.* Early Childhood Australia.

Edwards, S., & Cutter-Mackenzie, A. (2011). Environmentalising early childhood education curriculum through pedagogies of play. *Australasian Journal of Early Childhood* [E], *36*(1), 51–59.

Elliott, S., Ärlemalm-Hagsér, E., & Davis, J. (Eds.). (2020). *Researching early childhood education for sustainability: Challenging assumptions and orthodoxies* (Vol. 2). Routledge.

Elliott, S., & Davis, J. (2009). Exploring the resistance: An Australian perspective on educating for sustainability in early childhood. *International Journal of Early Childhood, 41*(2), 65–77.

Elliott, S., & Davis, J. (2023). *Early childhood Australia essentials: Sustainability in early childhood.* Early Childhood Australia.

Elliott, S., Edwards, S., Davis, J., & Cutter-McKenzie, A. (Eds.). (2013). *Best of sustainability: Research, theory and practice.* Early Childhood Australia.

Elliott, S., & Emmett, S. (1991/1997). *Snails live in houses too: Environmental education for the early years.* Martin Educational/RMIT Publishing.

Elliott, S., McCrea, N., Newsome, L., & Gaul, J. (2016). *Examining environmental education in NSW early childhood education services: A literature review with findings from the field.* State of NSW & Office of Environment and Heritage. https://www.environment.nsw.gov.au/resources/grants/160418-Early-Childhood-Report.pdf

Emery, S., et al. (2017). Transnational dialogues for sustainability research in early childhood education: A model for building capacity for ESD in universities? In W. L. Filho, C. Skanavis, A. do Paço, J. Rogers, O. Kuznetsova, & P. Castro, *Handbook of theory and practice of sustainable development in higher education* (pp.143–156). Springer. https://hdl.handle.net/102.100.100/23062802.v1

Engdahl, I., & Ärlemalm-Hagsér. (2020). *Att bli förskollärare: Mångfacetterad komplexitet* [Becoming a preschool teacher: Multifaceted complexity] (2nd ed.). Liber.

Engdahl, I., Pramling Samuelsson, I., & Ärlemalm-Hagsér, E. (2021). Swedish teachers in the process of implementing education for sustainability in early childhood education. *New Ideas in Child and Educational Psychology, 1*(1), 3–23.

Engdahl, I., Pramling Samuelsson, I., & Ärlemalm-Hagsér, E. (2023). Systematic child talks in early childhood education—A method for sustainability. *Children, 10*, 661. https://doi.org/10.3390/children10040661

EU Commission. (2022). *Council Recommendation on learning for environmental sustainability*. https://eur-lex.europa.eu/legal-content/EN/TXT/PDF/?uri=CELEX:52022SC0003&from=EN

Evans, N., Harwood, D., Furu, A.-C., & Alici, Ş. (2022). Challenges in reorienting early childhood education for sustainability in initial teacher education: Transforming, creating and expanding. *The Australian Journal of Teacher Education, 47*(5). https://doi.org/10.14221/ajte.2022v47n5.1

Freire, P. (1970). *Cultural action for freedom*. Harvard educational review.

Freire, P. (2021). *Pedagogy of hope: reliving pedagogy of the oppressed*. Bloomsbury Academic.

Fien, J. (1993). Education for the environment: Critical curriculum theorising and environmental education. *Deakin University*. https://doi.org/10.1016/S0140-6736(20)32290-X

Forsberg, B. (2002). *Lokal Agenda 21 för hållbar utveckling: en studie av miljöfrågan i tillväxtsamhället* [Local Agenda 21 for Sustainable Development—A Study of the Environmental Challenges in the Growth Society]. (Thesis). Umeå: Umeå universitet.

James, A., & Prout, A. (Eds.). (1990). *Constructing and reconstructing childhood*. Falmer Press.

Gaul, J., & Watson, D. (2019). *Eco Smart: A sustainable standard for early education*. Pademelon Press.

Gordon Community Children's Centre. (1993). *Playing for keeps: Environmentally responsible practices for early childhood learning*. Gordon Technical College.

Gustavsson, L., & Söderberg, L. (2021). *I ur och skur: upplevelsebaserat lärande för en hållbar livsstil i praktiken*[In ur och skur: experimental learning for a sustainable lifestyle]. The Swedish Outdoor Promotion Association, Friluftsfrämjandet, I ur och skur utveckling AB.

Hadley, F. et al. (2021). *Discussion paper: 2021 National Quality Framework Approved Learning Frameworks Update*. ACECQA.

Hatje, A. (1999). Från treklang till triangeldrama: Barnträdgården som ett kvinnligt samhällsprojekt under 1880–1940-talen. [*From trills to triangle drama: The Kindergarten as a Women's Society Project in the 1880s–1940s*]. Historiska media.

Harwood, D., Whitty, P., Green, C., & Elliot, E. (2020). Unsettling settlers' ideas of land and relearning land with Indigenous ways of knowing in ECEfS. In S. Elliott, E. Ärlemalm-Hagsér, & J. Davis (Eds.), *Researching early childhood education for sustainability: Challenging assumptions and orthodoxies* (pp. 25–37). Routledge.

Hedefalk, M., Almqvist, J., & Östman, L. (2014). Education for sustainable development in early childhood education: A review of the research literature. *Environmental Education Research, 21*(7), 975–990. https://doi.org/10.1080/13504622.2014.971716

Holmström, P., Bergenzaun, L., & Fernandez, H. (1992). *Låt det gå runt: Som vi gör - gör barnen* [Let it go around: As we do - do the children]. Bohusläns museum.

Immig, J. (2000). *The toxic playground: A guide to reducing chemical load in schools and childcare centres*. Total Environment Centre.

Intergovernmental Panel on Climate Change [IPCC]. (2023). *AR6 Synthesis Report: Climate Change 2023*. https://www.ipcc.ch/2023/

Keep Sweden Tidy Foundation. (2023). https://hsr.se/keep-sweden-tidy-foundation

Kinsella, R. (2007). *Greening services: Practical sustainability*. ECA.

Larsson, J., & Pramling Samuelsson, I. (2019). Collective resources as a precursor for educating children toward a sustainable global world. *ECNU Review of Education*, 2(4), 396–420. https://doi.org/10.1177/2096531119886506

Martin Korpi, B. (2016). *The politics of pre-school: Intentions and decisions underlying the emergence and growth of the Swedish preschool*. (4th ed.). Ministry of Education and Research.

Mulà, I., & Tilbury, D. (2023). Teacher education for the green transition and sustainable development. *EENEE Analytical Report*. https://doi.org/10.2766/144189

New South Wales Environment Protection Authority (NSW EPA). (2003). *Patches of green: A review of early childhood environmental education*. NSW EPA.

New South Wales Environmental Protection Agency (NSWEPA). (2003). *Patches of green early childhood environmental education in Australia: Scope, status and direction.* www.environment.nsw.gov.au/education/patchesofgreen.htm

New South Wales Early Childhood Environmental Education Network (NSWECEEN). (2013). *Evaluative report to ET: Leading sustainability in early childhood.* NSW ECEEN.

National Board of Health & Welfare. (1987). *Pedagogical Program for Preschool.* Swedish National Board of Health & Welfare.

Nikiforidou, Z., LavinyMiles, Z., & Luff, P. (2019). Rating education for sustainable development in the early Years: A necessity or a challenge? In Teacher Education for Sustainable Development and Global Citizenship Routledge, https://doi.org/10.4324/9780429427053-15

O'Gorman, L. (2017). Sustainability, the arts and big numbers: The challenge of researching children's responses to Chris Jordan's images. *International Journal of Early Childhood, 49*(3), 321–332.

OMEP. (2019). *ESD Rating scale* (2nd ed.) https://omepworld.org/education-3/

Piaget, J. (1953). *The origin of intelligence in the child.* Routledge & Kegan Paul.

Piaget, J., & Inhelder, B. (1969). *The psychology of the child.* Routledge & K. Paul.

Pramling Samuelsson, I., & Kaga, Y. (Eds.). (2008). *Early childhood and its contribution to a sustainable society.* UNESCO.

Pramling Samuelsson, I., Ärlemalm-Hagsér, E., Engdahl, I., Larsson, J., & Borg, F. (2021). Förskolans arbete med hållbarhet. [*Preschools' work with sustainability*]. Studentlitteratur.

Siraj-Blatchford, J., Mogharreban, C., & Park, E. (2016). *International research on education for sustainable development in early childhood.* Springer.

Sneddon, S., & Pettit, A. (2016). *Sustainability in action: Understanding sustainability and successfully incorporating it into everyday learning and practices.* Teaching Solutions.

Somerville, C., & Williams, M. (2015). Sustainability education in early childhood: An updated review of research in the field. *Contemporary Issues in Early Childhood, 16*(2), 102–117. https://doi.org/10.1177/1463949115585658

Sonter, L. et al. (2024). Exploring the Sustainable Development Goals: Beyond the environmental dimension. In J. Davis & S. Elliott (Eds.), *Young children and the environment.* Cambridge.

Stevenson, R. B., Brody, M., Dillon, J., & Wals, A. E. J. (Eds.). (2012). *International handbook of research on environmental education.* Routledge.

SOU. (1972:26). *Förskolan: betänkande* [The Commission on Nursery Provision]. Liberförlag/Allmänna förlaget.

SOU. (1997:157). *Att erövra omvärlden: förslag till läroplan för förskolan: Slutbetänkande* [Proposed curriculum for the Swedish preschool: Final report]. Fritze.

SOU. (2019:13). *Agenda 2030 and Sweden: Challenges and possibilities for the Earth.* The Swedish Government. https://www.government.se/legal-documents/2019/03/sou-201913/

Swedish National Agency for Education. (1998). *Curriculum for the preschool Lpfö98.*

Swedish National Agency for Education. (2018). *Curriculum for the preschool Lpfö18.* https://www.skolverket.se/publikationsserier/styrdokument/2018/laroplan-for-forskolan-lpfo-18

Swedish National Agency for Education. (2019). *Curriculum for the preschool Lpfö18.* Translated to English. https://www.skolverket.se/publikationer?id=4049

Technical & Further Education (TAFE) New South Wales Training & Education Support Industry Skills Unit. (2010). *Small footprints: CHCPR515A Develop and implement a program to support sustainable practice.* TAFE NSW Training & Education Support Industry Skills Unit.

The International Commission on the Futures of Education. (2021). *Futures of education: Learning to become.* UNESCO.

Tilbury, D., Coleman, V., & Garlick, D. (2005). *A national review of environmental education and its contribution to sustainability in Australia: School education.* Australian Government Department of the Environment & Heritage & Australian Research Institute in Education for Sustainability.

United Nations (UN). (2015). *Transforming our world: The 2030 agenda for sustainable development.* http://www.un.org/ga/search/view_doc.asp?symbol=A/RES/70/1&Lang=E

United Nations Educational, Scientific and Cultural Organisation (UNESCO). (2021). *Re-imagining our futures together: A new social contract for education*. UNESCO.

United Nations Educational, Scientific and Cultural Organisation (UNESCO). (2005). Decade of education for sustainable development 2005–2014. *Draft International Implementation Scheme*. UNESCO.

United Nations Educational, Scientific and Cultural Organisation (UNESCO). (2008). *The Gothenburg recommendations on education for sustainable development*. https://www.researchgate.net/publication/282121662_The_Gothenburg_Recommendations_on_Education_for_Sustainable_Development

United Nations Educational, Scientific and Cultural Organisation (UNESCO). (2012). *Education for sustainable development: Good practices in early childhood*. UNESCO. https://unesdoc.unesco.org/ark:/48223/pf0000217413

Vaealiki, S., & Mackey, G. (2008). Ripples of action: Strengthening environmental competency. *Early Childhood Folio, 12*, 7–11.

Victorian Environmental Education Council. (1992). *Learning to care for our environment: Victoria's Environmental Education Strategy*. Victorian Environmental Education Council.

Weldemariam, K., Boyd, D., Hirst, N., Sageidet, B. M., Browder, J. K., Grogan, L., & Hughes, F. (2017). A critical analysis of concepts associated with sustainability in early childhood curriculum frameworks across five national contexts. *International Journal of Early Childhood, 49*(3), 333–351. https://doi.org/10.1007/s13158-017-0202-8

Wenger, E. (2006). *Communities of practice: A brief introduction*. https://scholarsbank.uoregon.edu/xmlui/handle/1794/11736

Wilson, R. (1993). *Fostering a sense of wonder during the early childhood years*. Greyden Press.

Wilson, R. (1994). *Environmental Education at the Early Childhood Level*. NAAEE.

Wilson, R. (1995). Environmentally appropriate practices. *Early Childhood Education Journal, 23*, 107–110.

Wilson, R. (1996). The development of the ecological self. *Early Childhood Education Journal, 24*(2), 121–123.

World Commission on Environment and Development (WCED). (1987). *The Brundtland report: Our common future*. Oxford University Press.

Young, T., & Elliott, S. (2014). *Ways of thinking, acting and relating about sustainability*. Deakin West, ACT: Early Childhood Australia.

Chapter 3
Changing Theories and Pedagogies in ECEfS

Abstract In this chapter, the traces of changing theories and pedagogies across ECE and the EE/EfS/ESD fields—from the seventeenth to the twenty-first centuries—are outlined. These are changes that have led to current understandings of ECEfS that rethink the role of ECE as positioning the rights of children to be agents of change in challenging and troubling times. Further, the shift from the Anthropocene to the Ecocene is discussed as well as how transformative EC pedagogies can respond—globally and locally—to climate change, inequalities, and inequities.

3.1 Introduction

In this chapter, we recognise the traces of change in theories and pedagogies across the broader EC and EfS/EE/ESD fields that have led to the present constructions of ECEfS. Influential theoretical drivers include the Sociology of Childhood that has changed how people see children's capabilities and agency, systems thinking that seeks to overcome the isolation and fragmentation of individualistic approaches to sustainability, and the 'posties', e.g. post-colonialism, post-modernism, post-materialism, and post-humanism.

Similarly, EfS and ECEfS pedagogies have shifted focus from nature play/nature education offering intense personal connections to/with/in nature (Carson, 1962/2000; Naess, 1989) through to action-oriented EE and transformative educational practices underpinned by critical theories. These latter approaches align with rethinking of nature and environmental learning into the late twentieth century mantra of education 'in, about and for' the environment. Recent theorising such as post-humanism, post-Anthropocene, and new materialism offers further possibilities for rethinking ECEfS.

J. M. Davis et al., *Early Childhood Education for Sustainability*,
SpringerBriefs in Education, https://doi.org/10.1007/978-981-97-8405-9_3

3.2 Changing Ideas and Theoretical Shifts About Children and Childhood

In the following paragraphs, we chart thinking about children and childhood and key theoretical shifts over the centuries pertinent to our discussions about the history of ECEfS. We begin with developmentalism and move towards contemporary theorising in an era of climate crisis when children have most to lose and potentially gain in relation to their futures.

3.2.1 Societies in Transition: The Seventeenth and Eighteenth Centuries

How we understand the child and children in ECE cannot be separated from its history. The reproduction and reconstruction of childhood is shaped by societal, social, political, and scientific changes (Cunningham, 2005). In this history, or rather in these stories about ECE history, one persistent picture of ECE emerges of ongoing child development and as a future-oriented project, where the quest for a better society becomes central (Ärlemalm-Hagsér, 2013). This notion of ECE as guiding the future development of the child is made visible in how a child is understood, and what a child can do, ideas that intimately link the present with past and future.

In this view, notions of children emerge as offering hope of the future, a future shaped in a specific context—ECE. Here, scenarios are continuously constructed and reconstructed in relation to prevailing discourses about knowledge and society and, simultaneously, are the carriers of foundational ideas drawn from the origins of ECE (Dahlberg & Lenz Taguchi, 1994). Developmentalist ways of thinking continue to influence contemporary ECE practice in organisation, working methods, and holistic views of the child. Further, ECE is where place and space (indoors and outdoors in nature), and materials (e.g. aesthetics, playing, singing, and gardening) have significance for child development and where content knowledge emerges, including for example, in current teaching and learning for sustainability.

The societies in which ECE was formally established well over 150 years ago were societies in transition. Shifts from agrarian to industrial societies were difficult and the problems facing the disintegrating society were diverse. Demographic changes with migration from rural areas to cities were creating new problems of poverty and class conflict. The power of church and religion was diminishing, patriarchal structures were questioned, and social problems and individual rights were put on the political agenda. Conflicts became a fact of life. New social movements such as the women's movement and labour and trade union movements were changing social landscapes in the battle for justice, equality and democracy. In these troubled times, a strong belief in the power of upbringing and education developed. Education was seen as a means of achieving both stability and change, reducing social inequalities to relieve social unrest, and for raising the population's knowledge base. The possibility of education to create a new and better future was reinforced in these liberal-social ideas. Children were incorporated into this dream of a brighter future. Thus, ECE and early childhood teachers became part of this community building.

As Hultqvist (1990) commented, ECE developed as a social, political, and educational community project and, 'in the idea of the child, the ambition to contribute to the development of individuals conversing with the ambition to shape the new community' (p. 278).

3.2.2 Ideas About Children in the Enlightenment: The Seventeenth Century

Aligned with the great social changes of the Enlightenment in the seventeenth and eighteenth centuries and the emergence of the 'rationality' of scientific thinking, the shifts of agrarian societies towards industrialisation and urbanisation, new streams of thought were developing about the human, the child, and about children's education. The voices of Rousseau (1712–1778), Froebel (1782–1852) and Kant (1724–1804) were raised in relation to a new civilised and modern human being. In this rhetoric, hopes for a new kind of human began to emerge—a human who, through freedom, autonomy, reason, and self-discipline, could live a good life and be of benefit to societal development.

From a developmentalist view, children were defined in terms of 'nature' and childhood had its own peculiar way of seeing, thinking and feeling. According to Cunningham (2005), the ideas of childhood were developed between 1500 and 1900, beginning with a secular view of children as not adults requiring education and protection. This was not new; Italian humanist Erasmus wrote in the 1520 s about children's upbringing and education. He believed that 'nature had implanted in children the seed of a desire for knowledge and a power of memory greater than at any other age … and you should mould it that it takes on the best possible character' (Cunningham, 2005, p. 43).

3.2.3 Ideas About Children in the Enlightenment: The Eighteenth Century

In Rousseau's (1762/1979) philosophy of education, the formation and making of reason was central, Rousseau believed that the first upbringing was the most important and questioned the authoritarian upbringing that predominated at the time. According to Rousseau, children should grow up in freedom. Development occurs through the children's own experiences, preferably with nature as a source of inspiration and knowledge. In these experiences, the child's abilities to master their existence are developed, and inner control is acquired through self-discipline. Rousseau's child is a child of freedom, essentialism and nature, a child with an inherent drive, who is also vulnerable and fragile.

3.2.4 Developmentalism as a Human Process in the Nineteenth Century

Froebel, often claimed as the founder of ECE, was inspired by Rousseau's thoughts on the good and the free child. Nature is close in both Rousseau's and Frobel's ideas of the child and childhood. In Froebel's (1826/1995) thinking, the world was understood to be of a spiritual nature. Best (2016) explained this as:

> Spiritual education on this account is not an aspect of the curriculum; it is not religious education, nor is it limited to the first 'S' in SMSC (spiritual, moral, social and cultural dimension of cross-curricular approaches). Rather, to educate the whole child, to facilitate the development of its body, mind, feelings, aesthetic awareness, morality and potential for imaginative and creative activity. (p. 281)

According to Froebel, children were like plants to be taken care of and nurtured in well adapted environments, where they could be developed by self-directed activities. He created specific materials (gifts), games and songs for children, to promote children's learning and progress teaching to meet the needs of the child. In Froebel's philosophy, children's connections to nature were crucial 'as the basis for an understanding of themselves and their unfolding place in the world. Here, their sensory experiences would lead to a wonderment and an aesthetic response to the beauty and design (for all embodied mathematical order), and through reflection and meditation, a synthesis of the contrasts and opposites (the Seasons, life and death, night and day)' (Best, 2016, p. 278).

Rousseau and Froebel were also interested in the child's relationship with society and their social development. According to Rousseau, human freedom has limits. What can change this is if people's dependence on each other ceases, and a shared aspiration for the common good is established. In these early texts on children, childhood and upbringing, a liberal agenda emerged, in which human reason, freedom, autonomy and self-discipline were emphasised. Firstly, it was about the child as an individual and the development required to become a perfect human being. Secondly, this person had an important role in society and ongoing social development. However, this was an androcentric perspective promoting a masculine viewpoint in Rousseau and Froebel's thinking. For example, Rousseau's educational philosophy offers the rearing of the boy Emilé (Rousseau, 1762) with ideas that a man is civilised, rational, self-directed and self-governing, whereas the female does not hold these attributes.

3.2.5 The Next Phase: The Twentieth Century

Early in the twentieth century, new scientific theories about children's upbringing and development were being developed through the emergence of pedagogy and psychology as discipline areas. In Sweden, Alva Myrdal (1935) became an important voice as she advocated for child psychology research and for a scientific basis around children's education and development. Gesell's (1880–1961), developmental theory of the child's maturity, Piaget's (1896–1980) cognitive theory and Eriksson's (1902–1994) theory of the child's emotional development, from then on, had a dominant role in the understanding of children and childhood (Flewitt & Ang, 2020). During this period, Vygotsky (1896–1934) broadened developmentalist theorising by recognising that children's development occurs in social contexts and proposing his socio-cultural theory (Vygotsky, 1978). From this period, the idea of a universal and biologically predestined child was reinforced, with such theories informing ECE teacher education and practice. Developmental theories continue to be a reference point in contemporary ECE, even though more recent theorising questions such strongly developmentalist approaches.

3.2.6 Contemporary Discourses: The Twenty-First Century

A new time arrived with the twenty-first century, challenging these thought structures with individuality and pluralism. In the 1990s, children and childhood were described within the emerging childhood sociology in terms of social and cultural constructions that change over time (James et al., 1998). In addition to children and childhood, dichotomies such as nature/culture, passive/active, object/subject, dependent/independent, incompetent/competent, and being/prospective (Prout, 2005) were problematised. Previous understandings of children as development projects were challenged in relation to these dichotomies and childhood as a journey towards an aspiring adult questioned. Lee (2001) problematised positions such as being in the present (being) and prospective (becoming) and suggested these terms might be attributed not only to children, but also to adults, since being a child and being an adult, both imply constant change. This idea of human development can now be seen as part of the development of newer critical and feminist research perspectives. Children and childhood also began to be problematised in relation to other social power orders such as gender (Alanen, 1994), class and ethnicity (James et al., 1998). Topics central to the current century include children's rights and agency, post-colonialism, post-humanism, and new materialism influences, plus shifting from the Anthropocene to the Ecocene.

3.2.6.1 Children's Rights and Agency

One concept in use from the 1990s onwards is agency. Children are increasingly seen not as passive objects and recipients of knowledge, but as active and acting subjects who influence their surroundings through their encounters with individuals and the environment (James & James, 2004). Children's experiences became important perspectives in contemporary childhood research (Christensen & James, 2000).

In parallel with these newer trends in child and childhood research, political discussion developed around children's rights, the right to make their voices heard and have influence over issues concerning their futures and contexts. In the political arena, the UN Convention on the Rights of the Child (1989) was ratified, underpinned by the idea of a pluralistic child with rights as societal actors. According to Quennerstedt (2011), the political construction of children's rights in relation to education are now embedded in specific national, cultural, political, and economic circumstances, and further elaborated by Visnjić-Jevtić et al. (2021). Thus, the child who, at the beginning of the twentieth century, became the subject of a reform project in the struggle for morality during the disintegration and crisis of society, appears at the end of the twentieth century, in a social project for sustainable futures with children as significant actors.

Identifying children as important actors with capabilities and in their own lives is not a new phenomenon, as Rousseau (1762/1892) wrote. However, over the years, children's abilities to be actors and recognised as agents of change in their relationships with adults have been mixed. In the present day, the rhetoric of child competence and children's rights has more strongly emerged. This changing approach to children and childhood has been made visible with Davis's (2015) theoretical starting points for children's sustainability learning, paraphrased as having a:

- rights dimension—children have the right to express their opinions and to make their voices heard;
- capabilities dimension—children are capable of taking part in and shaping their world; and
- participatory actor dimension—children are social actors with abilities to challenge and act.

Davis argues that children should have the opportunity, together with adults, to challenge and actively change unsustainable thinking and action at the structural and partisan level, in ECE and broader society. Active citizenship includes having opportunities to engage in democratic processes with the aim of discussing and problematising issues around environmental concerns, inequality and the meanings of citizenship.

3.2.7 *Post-colonialism Influences: Twentieth and Twenty-First Centuries*

In 2004, Cannella and Viruru, by using post-colonial cultural theories, have also challenged the oppressive practices of dominant Western discourses about children and the ways in which majority world ideas about educating became a powerful discourse on how to educate children. Also, Gupta (2013) raised questions about the risk of cultural 'incursion' as dominant Western early childhood pedagogies and ideologies contributed to globalisation and change in ECE policies internationally. Her study identified how Western narratives promote child-centred and play-based curricula, a significant shift from the traditional teacher-directed education of countries such as India, Singapore, and China. She stressed that 'Western concepts cannot be readily and easily implemented within the practical context of culturally diverse worldviews' (Gupta, 2013, p. 223) and argues for a 'pedagogy of the third space'. This is a pedagogical perspective that holds alternative practical and theoretical interpretations and opens possibilities to acknowledge both local and global contexts. Often referred to as 'glocal', there is potential for the coexistence of diverse educational approaches derived from both global discourses and local narratives. Also, Cannella (1997) raised the power discourses in society and education, specifically noting there are particular groups of children who are more privileged than others, and inequality needs to be discussed in relation to ECE.

3.2.8 *Post-modernity, Post-humanism, and New Materialism Influences into Contemporary Times*

As we have illustrated above, over past centuries, ideas about educating children have changed significantly. Nevertheless, ECE remains one of the keys to ensuring that citizens develop socially appropriate competencies and skills for participating in increasingly complex societies. Thus, earlier constructs have been challenged by a suite of newer theoretical perspectives such as post-humanism and new materialism that emerged around the 1990s (Moss, 2019; Taylor & Hughes, 2016). Post-humanism attempts to dissolve dualisms, such as nature–culture, in children's development (both nature and culture) and in relationships between the human–non-human. For example, Deleuze and Guattari's (2004) notion of rhizomatic thinking challenges earlier understandings of learning and education by exploring notions of the unexpected, multiplicity and assemblage.

Recently too, research challenging dominant anthropocentric understandings has explored alternative ways of understanding and promoting sustainability in ECE (Cutter-Mackenzie-Knowles et al., 2020; Malone, 2016; Rooney & Blaise, 2023; Taylor, 2016; Weldemariam, 2017). These studies, especially, emphasise children's experiences in nature, to develop knowledge of nature, and the necessity for building respectful relationships with nature and the non-human world, aspects crucial for understanding human intertwining with nature and promoting sustainable futures. In relation to new materialism, Lenz Taguchi (2010) employed the concept of intra-action pedagogy, and stressed a shift from giving attention to the individual and groups of individuals to a focus on the intra-active relationships between all living organisms and the material environment. This stance challenges the common ECE focus on children's agency and instead offers opportunities to deeply explore relationalities. Common and entangled human and more-than-human worlds are envisaged in post-humanism, driven by ethical relations and a sense of collectivism.

3.3 Shifting from the Anthropocene to the Ecocene

More recently, the Anthropocene as a concept has begun to appear in ECE and sustainability research (Somerville & Williams, 2015). A significant article is Steffen et al. (2007) where they argued that the Anthropocene, as a geological epoch, should succeed the Holocene. They showed that human impacts on the Earth and the atmosphere are now clearly visible through climate warming, loss of biodiversity and habitats, rising sea levels and water, soil and air pollution. These changes generate risks beyond all previous human experiences and, in the long run, challenge the Earth's nature and societies, globally and locally, risks that are ever-increasing and pose a critical threat to all the world's children, if no radical shifts occur (Currie & Deschênes, 2016). Küpers (2020) argues for a shift from Anthropocene to Ecocene by proposing an embodied perspective 'where cultivating a non-anthropocentric ethos … that allows human and non-or other-than-human worlds to peacefully coexist' (p. 14). The global challenge is to actively work towards the next epoch, the Ecocene, as ecologically informed living and this will require societal shifts at all levels (Boehnert, 2018), critically beginning in early childhood.

3.4 Changing Ideas About Pedagogies

Aligned with changing ideas about children and childhoods and the theoretical shifts outlined above, we now focus on pedagogies and shifting pedagogical approaches evident over time in early childhood EE/ESD/EfS. In particular, there has been a shift towards intense personal connections with Nature, referred to as Deep Ecology.

The interconnections between humans and nature and the need for critical awareness about human behaviour in relation to the environment began in the 1960s with Rachel Carson (1965/1998), biologist and environmentalist, and her book *Silent Spring*. Carson argued that widespread chemical use, developed by expanding chemical industries and utilised extensively by agricultural communities to prevent insects destroying crops, was poisoning the environment. Further, she maintained that chemical pollution had reached the point where global ecological well-being was at risk.

Her book provoked public awareness and debate about pesticide danger, leading to changes in US government policies and inspired the modern ecological movement.

Carson (1965/1998) argued that children meet nature through all their senses with wonder and fascination, a sensibility that later in life is easily lost if it is not cared for. She wrote in *The Sense of Wonder* about the importance of nature encounters in childhood. What is crucial, Carson believed, was to develop a sense of wonder, described as more important than developing knowledge about nature. The deep feeling for nature is a recurring theme in the discussion of human relationships with nature.

One of the important philosophers was Arne Naess, a Norwegian philosophy professor. Naess (1989) began 'deep ecology', an environmental movement that values all living things on the basis that they have equal value and everything in life is connected. He identified that the impact of humans on the biosphere, for example, is fundamentally a philosophical question about life. This ecophilosophy, offers a platform of values including: there are intrinsic values in both human and non-human life on Earth; the richness and diversity of life forms have value in themselves and contribute to the flourishing of human and non-human life on Earth; and humans have no right to reduce this richness and diversity except to satisfy vital needs.

Naess challenges anthropocentric thinking by taking a deep ecological perspective (referred to as 'deep green'), and he chooses a pluralistic path of action and change. First, it is the structures within the 'system' that should change—views of human and nature, human relationships with the Earth, and the necessary shift from consumeristic and materialistic lifestyles—to increased focus on quality of life. Second, it is the change of the 'system' itself that is most important by restructuring the economic, technological, and ideological foundations of societies, aiming to support environmental and sustainable development. Naess' thoughts have inspired various ideas that continue to unfold in EfS today challenging dominant anthropocentric understandings in EfS pedagogical studies, for example, Cutter-Mackenzie-Knowles et al. (2020), Rooney and Blaise (2023), Taylor (2016), and Weldemariam (2017) as mentioned earlier.

3.4.1 Education In, About and For the Environment

Along with the historical importance of nature play in ECE discussed earlier, Naess' ideas helped deepen education 'in the environment' in the 1970s and 1980s. A new way of looking at EE began to emerge in the 1990s that became known as education in, about and for the environment (Fien, 1993). Davis (1996) applied the organiser of 'in, about and for the environment' to ECE, in an early practitioner article paraphrased below. Education in the environment, she wrote, provides for direct experience with natural environments, seeking to develop positive feelings and attitudes for natural systems, to foster empathy for others—human and non-human—and to explore in a practical sense environmental conflicts and issues. Encouraging children to feel comfortable in the outdoors is a primary pedagogical practice, where they can 'ooh' and 'ahh' a lot at what they see and do outdoors.

Davis wrote that education about the environment seeks to provide learners with understandings about how natural systems work, appreciate their complexity, and recognise how these and human systems interact. Foci for learning included principles of the water cycle, oxygen and carbon cycles, food chains, and food webs. Today, understanding climate systems and human interactions impacting these systems would be seen as crucial. Pedagogies about the environment might include water conservation, planting vegetation to improve biodiversity, and teaching children not to take bucketloads of 'treasures' from the beach.

Lastly, in her 1996 article, Davis wrote that education 'for' the environment was concerned with social critique of existing ways of interacting with the environment and taking social action for change. There was explicit emphasis on working together for the common good—rather than enhancing individualistic approaches to environmental matters—often seen as underpinning in and about approaches—that differentiates education for the environment from the other two forms. Examples might include examining unchecked water play from the perspective of Australian children growing up in the driest inhabited continent on the globe. From this summary, it is evident an early transition was beginning to occur—in Australia at least—towards a more activist form of EE in ECE. Especially in schools, education 'for' the environment became a dominant form of EE as classes and schools worked on action projects such as tree-planting/habitat restoration and waste reduction in schools, often with a whole school emphasis (see Chap. 2. for the Swedish historical context).

3.4.2 *Education for Sustainability: Systems, Action, and Participation*

In the 2000s, the language around EE began to shift towards EfS/ESD terminology, representing a stronger global emphasis on a multidimensional approach to sustainability—more than the existing focus primarily on 'environmental concerns'. EfS aimed to generate a much more participatory approach to learning and teaching about environmental/sustainability concerns, where the educational setting was acknowledged as part of a wider 'system' rather than as a standalone entity. Hence, building connections with the local community, NGOs, local businesses, and government agencies such as health departments and environmental departments to encourage greater alignment with common goals became a feature. Further, EfS doubled down on the action-orientation of learning and teaching. Actions might range from children physically investigating a local biodiversity issue such as a weed-clogged creek, for example, and then planning action to remediate, and then carrying out these plans by de-weeding and revegetating the creek bank. However, actions may also venture into conducting local campaigns within a council/municipality area or even global campaigns such as the *School Kids for Climate*.

In the ECE sector, the Sustainable Planet Project at Campus Kindergarten in Brisbane Australia is seen as an exemplar, even today, of a participatory, action-oriented approach to ECEfS (Davis et al., 2005) where, among other things, centre children wrote to the Prime Minister of Australia seeking greater government effort on global warming. They also planned and actioned their own tree-planting program in the neighbourhood.

Another key aspect of EfS is enacting a whole school approach. In an Australian Research Institute in Education for Sustainability (ARIES) report, Henderson and Tilbury (2004) identified that school programs using the Whole Schools Approach (WSA) commenced in early 1990s. In the last ten years, WSA has been further developed (Mathie & Wals, 2022). This approach is built on the requirement for developing a broad framework for re-orienting education. The approach is holistic and involves all stakeholders within an educational institution such as a school or preschool to engage in meaningful, connected activities and actions to explore and address sustainability issues. These aspects are incorporated in Fig. 3.1, depicted as a flower with multiple flower petals.

Working with the WSA requires the involvement of all educational aspects (see Fig. 3.1) from management, to pedagogy, curriculum and the community in the broad work needed for progressing sustainability.

Fig. 3.1 Whole school approach (WSA) from Mathie and Wals (2022, p. 6) (designed by Coöperatie Leren voor Morgen, The Netherlands). Permission granted

Thus, a school or preschool that strives to work with sustainability needs to develop a holistic view where they integrate a long-term sustainability mindset: in management, the physical environment, teaching, and play. To work in this way, the whole institution/whole settings approach can also be envisaged as developing 'cultures of sustainability' (Davis, 2008).

3.5 Nature Play: Shifting Pedagogies

Children's connections with nature facilitated by playing in nature have long been an ECE tenet, harking back to the work of Frobel and Pestalozzi in the 1800s and the Scandinavian nature kindergartens of the 1950s (Williams-Siegfredsen, 2017). Fast forward to contemporary times when children's nature disconnect has been popularly described as 'nature deficit disorder' by Louv (2010) and nature play has become 'newly important' as a pedagogical tool (Gray, 2018). Such calls for reconnecting with nature have been realised in a significant increase in the diversity and number of programs that invite children and educators to explore and learn outdoors in, about and with nature. For example, regular excursion type programs beyond the confines of highly denatured manufactured playspaces are now highly evident internationally, from the forest schools of the United Kingdom to the nature preschools of the USA and beach or bush kinders of Australia (Hughes et al., 2022; Knight, 2013; Natural Start Alliance (NSA), 2022). In addition to these programs, there have also been shifts in the design of school or centre based playgrounds to incorporate natural elements (Gamson-Danks, 2010; Little et al., 2017) and this is now increasingly reflected in practice, more so in the ECE sector than schools as yet. Whether an excursion type program or on-site re-naturing of playspaces, research extols the learning and development benefits children may gain from playing in nature (Dankiw et al., 2020; Mygind et al., 2018). The reported benefits are diverse ranging from enhanced physical development (McClain & Vandermaas-Peeler, 2016), to the promotion of children's executive functioning (Carr et al., 2017) and positive risk management supporting children's health and well-being (Sandseter & Kleppe, 2019). In this context, often informed by children's health and well-being concerns compounded by the impacts of climate change, naturalised playspaces offer a breadth of provocations for learning.

Educator rethinking about their pedagogical approaches is most critical to capture and share teachable moments in dynamic natural settings. There are many possibilities from the sensory awe and wonder of being in nature, originally highlighted by Carson (1965/1998), to the explanatory power of scientific learning about nature and more recently, post-humanist theorising about ethically being with nature to challenge the pervasive anthropocentric lens that drives climate change (Cutter-Mackenzie-Knowles et al., 2020; Nelson & Hodgins, 2020; Rooney & Blaise, 2023). Our experience suggests educators may default to the more familiar sensory or scientific explorations with children (Elliott & Young, 2015), rather than tackle the ethical stances and global eco-centric lens required for authentically being with nature. We identify this as a pivot point for educators in rethinking their pedagogical interactions and there are many past and current calls for critically rethinking pedagogy in the face of climate change (Ärlemalm-Hagsér, 2013; International Commission on the Futures of Education, 2021; Kelly & White, 2013; Rooney & Blaise, 2023).

Bentsen et al. (2010, p. 236) have previously noted 'it is important to be aware of local, regional and national contexts, and of how curriculum and outdoor educational practices are framed and shaped by cultural, social, political and geographical factors'. These factors are rapidly changing, and recent research highlights the potential for deeper interrogation of relationships with others, both human and more than human, in natural settings with children (Harwood et al., 2019; Rooney & Blaise, 2023). Educators must be responsive and innovative in these settings and recognise both their intergenerational and interspecies responsibilities for shifting towards sustainable futures for all.

3.6 Transformative Pedagogies Underpinned by Critical Theory

Jickling and Sterling (2017) and Jickling (2017) argue that unthinking towards sustainable futures will not happen by itself; an education that questions and interferes with unsustainable ways of thinking and acting is essential. They argue for transformative pedagogies where EfS is imaginative, creative, courageous, and radical (Jickling & Sterling, 2017)—an urgent commitment to reducing the negative effects of human impact on the Earth (Jickling & Sterling, 2017). One way for creating change is to equip children and young people with competencies for critical and ethical approaches to action and the knowledge to respond to complex and difficult environmental, economic, and social challenges. This is an education where human interconnectedness with nature is in focus and alongside the requirement for humans to fully understand their impact on the Earth and atmosphere (Kopnina, 2020). Kopnina (2020) emphasises the development of non-Anthropocentric ethics, a planetary ethic, and degrowth as key strategies for addressing unsustainability. Jickling (2017) suggests the term 'post-sustainability' and argues for educational and pedagogical approaches to be challenged, disruptive and transformative for overcoming the challenges increasingly prominent in the Anthropocene. Säfström and Östman (2020) concur, underlining the necessity for teaching characterised by being transactive, where children and young people are active in the world and where every day ethical dilemmas create opportunities to challenge values and generate new or revised values.

3.7 Conclusion

In this chapter, we have outlined the traces of change in theories and pedagogies—from the seventeenth to the twenty-first centuries—across ECE and EE/EfS/ESD leading to current ECEfS constructions. As described here, theoretical and pedagogical understandings have shifted over the years positioning children at the forefront of changes that require rethinking how humans respond to climate change, inequalities and inequities, globally and locally. Through interrogating these theories and pedagogies, we also acknowledge the need for the ongoing development of critical and analytical understandings of the global and local, 'glocal' as Gupta suggests (Gupta, 2013). Currently, a range of newer theoretical and pedagogical ideas have emerged, including: post-colonial pedagogies (Cannella & Viruru, 2004; Gupta, 2013); place pedagogies (Massey, 2013; Rooney & Blase, 2023; Taylor, 2016); post-humanist theory and pedagogies (Malone, 2016; Taylor & Hughes, 2016); post-structural feminist theories and pedagogies related to intra-action and materialism (Lenz Taguchi, 2010); indigenous theories and pedagogies (Miller, 2014; Ritchie, 2014); and theories and pedagogics of animals as co-creators in ECE (Born, 2018; Young & Bone, 2020).

It is anticipated that ECEfS will continue to evolve on multiple paths into a post-Anthropocentric and post-sustainability era (Jickling & Sterling, 2017,) as it moves towards the Ecocene (Boehnert, 2018; Küpers, 2020), with the potential to target non-Anthropocentric ethics, planetary ethics, and degrowth (Kopnina, 2020) as key strategies for addressing unsustainability in ECE. To conclude this chapter, we acknowledge that we have been unable to do justice to all the research, researchers, and the full range of theoretical and pedagogical perspectives developed in the last decade, but we do hope that we have been able to provide readers with some sense of the vibrant and expanding research quilt that is shaping and reshaping ECEfS.

References

Alanen, L., et al. (1994). Gender and generation: Feminism and the child question. In J. Qvortrup (Ed.), *Childhood matters: Social theory, practice and politics* (pp. 27–42). Avebury.

Ärlemalm-Hagsér, E. (2013). Respect for nature: A prescription for developing environmental awareness in preschool. *Centre for Educational Policy Studies Journal, 3*(1), 25–44.

Bentsen, P., Jensen, F., Mygind, E., & Randrup, T. (2010). The extent and dissemination of udeskole in Danish schools. *Urban Forestry & Urban Greening, 9*(3), 235–243. https://doi.org/10.1016/j.ufug.2010.02.001

Best, R. (2016). Exploring the spiritual in the pedagogy of Friedrich Froebel. *International Journal of Children's Spirituality, 21*(3–4), 272–282. https://doi.org/10.1080/1364436X.2016.1231664

Boehnert, J. (2018). *Design, ecology, politics: Towards the Ecocene*. Bloomsbury.

Born, P. (2018). Regarding animals: A perspective on the importance of animals in early childhood environmental education. *International Journal of Early Childhood Environmental Education, 5*(2), 46–57.

Cannella, G. (1997). *Deconstructing early childhood education: Social justice and revolution*. Peter Lang.

Cannella, G., & Viruru, R. (2004). *Childhood and postcolonization: Power, education and contemporary practises*. Routledge/Falmer.

Carson, R. L. (1962/2000). *Silent spring*. Penguin Books.

Carson, R. L. (1965/1998). *The sense of wonder.* HarperCollins.

Carr, V., Brown, R., Schlembach, S., & Kochanowski, L. (2017). Nature by design: Playscape affordances support the use of executive function in preschoolers. *Children, Youth & Environments, 27*(2), 25–46.

Christensen, P., & James, A. (Eds.). (2000). *Research with children. Perspectives and practices.* Routledge.

Currie, J., & Deschenês, O. (2016). Children and climate change: Introducing the issue. *The Future of Children, 26*(1), 3–9. https://www.jstor.org/stable/43755227

Cunningham, H. (2005). *Children and childhood in western society since 1500* (2nd ed.). Pearson/Longman.

Cutter-Mackenzie-Knowles, A., Malone, K., & Barratt Hacking, E. (2020). Childhoodnature: An assemblage adventure. In A. Cutter-Mackenzie-Knowles, K. Malone, & E. Barratt Hacking (Eds.), *Research handbook on childhoodnature* (pp. 1–15). Springer. https://doi.org/10.1007/978-3-319-67286-1_2

Dahlberg, G., & Lenz Taguchi, H. (1994). *Förskola och skola: Om två skilda traditioner och om visionen om en mötesplats.* HLS förlag.

Dankiw, K., Tsiros, M., Baldock, K., & Kumar, S. (2020). The impacts of unstructured nature play on health in early childhood developments: A systematic review, *PLoS ONE, 15*(2). https://doi.org/10.1371/journal.pone.0229006

Davis, J. (1996). Early childhood environmental education. *Educating Young Children, 2*(4), 38–39.

Davis, J., Rowntree, N., Gibson, M., Pratt, R., & Eglington, A. (2005). Creating a culture of sustainability: From project to integrated education for sustainability at Campus Kindergarten. In W. L. Filho. (Ed.), *Handbook of Sustainability Research* (pp. 563–594). Peter Lang Publishing.

Davis, J. (2008). What might education for sustainability look like in early childhood? A case for participatory, whole-setting approaches. In Y. Kaga & I. Pramling Samuelsson. (Eds.), *The role of early childhood education for a sustainable society* (pp. 18–24). UNESCO.

Davis, J. (2015). *Young children and the environment: Early learning for sustainability* (2nd ed.). Cambridge University Press.

Deleuze, G., & Guattari, F. (2004). *A thousand plateaus: Capitalism and schizophrenia.* Continuum.

Elliott, S., & Young, T. (2015). Nature by default in early childhood education for sustainability. *Australian Journal of Environmental Education, 32*(1), 57–64. https://doi.org/10.1017/aee.2015.44

Fien, J. (1993). *Education for the environment: critical curriculum theorising and environmental education.* Deakin University.

Flewitt, R., & Ang, L. (2020). *Research methods for early childhood education.* Bloomsbury.

Gamson-Danks, S. (2010). *Asphalt to ecosystems: Design ideas for schoolyard transformation.* New Village Press.

Gray, T. (2018). Outdoor learning: Not new, just newly important. *Curriculum Perspectives, 38*, 145–149. https://doi.org/10.1007/s41297-018-0054-x

Gupta, A. (2013). Play - Early childhood pedagogies and policies in a globalizing Asia. In O. F. Lillemyr, S. Dockett, & B. Perry (Eds.), *Varied perspectives on play and learning. Theory and research on early years education* (pp. 213–230). Information Age Publishing.

Harwood, D., Barratt, J., & Collier, D. (2019). Entanglements in the forest: Children's embodiment of the orange GoPro camera. *International Journal of Early Childhood Environmental Education* —Special issue-Living within Precarious Times: Posthumanist Possibilities for Early Childhood Environmental Education, *7*(1), 57–72.

Henderson, K., & Tilbury, D. (2004). *Whole-school approaches to sustainability: An international review of sustainable school programs.* Report prepared by the Australian Research Institute in Education for Sustainability (ARIES) for The Department of the Environment and Heritage, Australian Government.

Hughes, F., Elliott, S., Anderson, K., & Chancellor, B. (2022). *Early years learning in Australian natural environments.*
Hultqvist, K. (1990). Förskolebarnet: en konstruktion för gemenskapen och den individuella frigörelsen: En nutidshistorisk studie om makt och kunskap i bilden av barnet i statliga utredningar om förskolan. [The preschool child: a construct for community and individual emancipation: A contemporary historical study of power and knowledge in the image of the child in government investigations of preschool]. Stockholm University.
International Commission on the Futures of Education. (2021). *Futures of education: Learning to become.* UNESCO. https://unesdoc.unesco.org/ark:/48223/pf0000375746/
James, A., & James, A. L. (2004). *Constructing childhood: Theory, policy and social practice.* Palgrave Macmillan.
James, A., Jenks, C., & Prout, A. (1998). *Theorizing childhood.* Polity Press.
Jickling, B. (2017). Education revisited: Creating educational experiences that are held, felt and disruptive. In B. Jickling & S. Sterling (Eds.), *Post-sustainability and environmental education: Remaking education for the future* (pp. 15–30). Palgrave Macmillan.
Jickling, B., & Sterling, S. (2017). Post-sustainability and environmental education: Framing issues. In B. Jickling, & S. Sterling (Eds.), *Post-sustainability and environmental education. Reimaging education for the future* (pp. 1–11). Palgrave Macmillan.
Kelly, J., & White, E. J. (2013). *The Ngahere Project: Teaching and learning possibilities in nature settings.* Wilf Malcolm Institute of Educational Research, University of Waikato. http://www.waikato.ac.nz/data/assets/pdf_file/0007/146176/Ngahere-project_3-2013-03-14.pdf
Knight, S. (2013). *International perspectives on forest school: Natural spaces to play and learn.* Sage.
Kopnina, H. (2020). Education for the future? Critical evaluation of education for sustainable development goals. *The Journal of Environmental Education, 51*(4), 280–291. https://doi.org/10.1080/00958964.2019.1710444
Küpers, W. M. (2020). From the Anthropocene to an "Ecocene"—eco-phenomenological perspectives on embodied, anthrodecentric transformations towards enlivening practices of organising sustainably. *Sustainability, 12*(9), 3633; https://doi.org/10.3390/su12093633
Lee, N. (2001). *Childhood and society: Growing up in an age of uncertainty.* Open University Press.
Lenz Taguchi, H. (2010). *Going beyond the theory/practice divide in early childhood education: Introducing an intra-active pedagogy.* Routledge.
Little, H., Elliott, S., & Wyver, S. (2017). *Outdoor learning environments: Spaces for exploration, discovery and risk taking in the early years.* Allen & Unwin.
Louv, R. (2010). *Last child in the woods: Saving our children from nature-deficit disorder.* Atlantic.
Malone, K. (2016). Reconsidering children's encounters with nature and place using posthumanism. *Australian Journal of Environmental Education, 32*(1), 1–15.
Massey, D. (2013). Space, place and gender. *WILEY.* https://doi.org/10.5749/j.cttttw2z
Mathie, R. G., & Wals, A. E. J. (2022). *Whole School Approaches to sustainability: Exemplary practices from around the world.* Education & Learning. https://lerenvoormorgen.org/wp-content/uploads/2022/04/WSA-REPORT-1-4-no-facing-pages_digital-web.pdf
McClain, C., & Vandermaas-Peeler, M. (2016). Social contexts of development in natural outdoor environments: Children's motor activities, personal challenges and peer interactions at the river and the creek. *Journal of Adventure Education and Outdoor Learning, 16*(1), 31–48. https://doi.org/10.1080/14729679.2015.1050682
Miller, M. (2014). Intercultural dialogues in early childhood education for sustainability: Embedding Indigenous perspectives. In J. Davis & S. Elliott (Eds.), *Research in early childhood education for sustainability: International perspectives and provocations* (pp. 63–78). Routledge.
Moss, P. (2019). *Alternative narratives in early childhood: An introduction for students and practitioners.* Routledge.
Mygind, L., Kjeldsted, E., Hartmeyer, R., Mygind, E., & Bentsen, P. (2018). A systematic review of the effects of immersive nature-experience on mental, physical and social

health. https://vide-nomfriluftsliv.dk/sites/default/files/pdf/Pdf-filer/lang_rapport_effekter_af_friluftsliv_p_sundheden.pdf

Myrdal, A. (1935). *Stadsbarn: En bok om deras fostran i storbarnkammare* [City Children: A Book of Their Nurturing in Large Nurseries]. Kooperativa förbundet.

Næss, A. (1989). *Ecology, community and lifestyle: Outline of an ecosophy*. Cambridge University Press.

Natural Start Alliance (NSA). (2022). *Nature preschools in the United States 2022 National Survey*. NSA.

Nelson, N. & Hodgins, B.D. (2020). Unruly voices: Growing climate action pedagogies with children and trees. In S. Elliott, E. Ärlemalm-Hagsér, & J. Davis (Eds.), *Researching early childhood education for sustainability: Challenging assumptions and orthodoxies* (Vol. 2, pp. 150–165). Routledge.

Prout, A. (2005). *The future of childhood*. Routledge/Falmer.

Quennerstedt, A. (2011). The political construction of children's rights in education—A comparative analysis of Sweden and New Zealand. *Education Inquiry, 2*(3), 453–471.

Ritchie, J. (2014). Learning from the wisdom of elders. In J. Davis & S. Elliott (Eds.), *Research in early childhood education for sustainability: International perspectives and provocations* (pp. 36–49). Routledge.

Rooney, T., & Blaise, M. (2023). *Rethinking environmental education in a climate change era: Weather learning in early childhood*. Routledge.

Rousseau, J. (1762). *Émile, ou, De l'éducation* [Emile, or On Education]. Francfort.

Säfström, C.-A., & Östman, L. (2020). Transactive teaching in a time of climate crisis. *Journal of Philosophy of Education, 54*(4), 989–1002. https://doi.org/10.1111/1467-9752.12477

Sandseter, E. E., & Kleppe, R. (2019). Outdoor risky play. In M. Brussoni (Ed.), *Encyclopedia on Early Childhood Development*. http://www.child-encyclopedia.com/outdoor-play/according-experts/outdoor-risky-play.

Somerville, M., & Williams, C. (2015). Sustainability education in early childhood: An updated review of research in the field. *Contemporary Issues in Early Childhood, 16*(2), 102–117. https://doi.org/10.1177/1463949115585658

Steffen, W., Crutzen, P. J., & McNeill, J. (2007). The Anthropocene: Are humans now overwhelming the great forces of nature? *Ambio, 36*(8), 614–621.

Taylor, A. (2016). Beyond stewardship: Common world pedagogies for the Anthropocene. *Environmental Education Research, 23*(10), 1448–1460.

Taylor, C. A., & Hughes, C. (Eds.). (2016). *Posthuman research practices in education*. Palgrave Macmillan.

Visnjić-Jevtić, A., Sadownik, A. R., & Engdahl, I. (2021). *Young children in the world and their rights: Thirty years with the United Nations Convention on the Rights of the Child* (Vol. 35). Springer International Publishing AG. https://doi.org/10.1007/978-3-030-68241-5

Vygotsky, L. (1978). *Mind in society: Development of higher psychological processes*. Harvard University Press.

Weldemariam, K. (2017). Challenging and expanding the notion of sustainability within early childhood education: Perspectives from post-humanism and/or new materialism. In O. Franck & C. Osbeck (Eds.), *Ethical literacies and education for sustainable development: Young people, subjectivity and democratic participation* (pp. 105–126). Palgrave Macmillan.

Williams-Siegfredsen, J. (2017). *Understanding the Danish Forest School approach* (2nd ed.). Routledge.

Young, T., & Bone, J. (2020). Troubling intersections of childhood/animals/education: Narratives of love, life and death. In A. Cutter-Mackenzie-Knowles, K. Malone, & E. Barratt Hacking (Eds.), *Research handbook on childhoodnature. Springer International handbooks of education*. Springer. https://doi.org/10.1007/978-3-319-67286-1_73

Chapter 4
Shifts Across ECEfS Research

Abstract In this chapter, we present an analysis of shifts and outputs across ECEfS research between 2014 and 2022, measured through journal papers, special journal issues, and research-based books in ECEfS published over recent decades. Key themes about ECEfS—emerging from the analysis of the 108 sampled articles—are presented and discussed. Additionally, we highlight some less prominent themes we believe require extra attention in future ECEfS research.

4.1 Introduction

In addition to the shifts outlined in previous chapters, significant changes have occurred across research. An encouraging ECEfS research trajectory has emerged since 2009, with shifts evident away from a mostly nature/environmental and psychological well-being focus, through to a diverse body of research and researchers producing multiple studies. Currently, there are independent research groups in various parts of the world conducting literature reviews and studies around early childhood EE/EfS/ESD. As noted in Chaps. 2 and 3, the TND network has been instrumental to this research trajectory, building international partnerships and research collaborations (Dean & Elliott, 2022; Emery et al., 2017). From our perspectives, this newer research and theorising has inspired these highlights:

- reimagining children's rights towards an expanded view incorporating environmental/sustainability dimensions (Davis, 2014);
- examining young children's understandings of sustainability (Christensen, 2021);
- rethinking Bronfenbrenner's anthropocentric model of human development (Elliott & Davis, 2018);
- recognising First Nations Peoples worldviews (Harwood et al., 2020; Miller, 2014; Ritchie, 2014) and non-Western approaches to ECEfS (Almeida & O'Hara, 2020; Inoue, 2014);
- provoking deeper EfS thinking and practice with Swedish in-service preschool teachers (Ärlemalm-Hagsér & Pramling Samuelsson, 2021; Engdahl et al., 2023);

J. M. Davis et al., *Early Childhood Education for Sustainability*,
SpringerBriefs in Education, https://doi.org/10.1007/978-981-97-8405-9_4

- expanding research on the dimensions of sustainability beyond the environmental, (cultural, Kim & Dreamson, 2020; social, Paujik et al., 2021; economic, Ärlemalm-Hagsér et al., 2018; Borg, 2017); and
- investigating nature play to critically rethink human/nature dichotomies and children's agency (Elliott & Davis, 2018; Elliott & Young, 2016; Hughes et al., 2022; Nelson & Hodgins, 2020; Weldemariam & Wals, 2020).

Additionally, we consider here where ECEfS research is cited beyond the expected ECE and EfS journals and texts. For example, growing numbers of such works are regularly cited in fields including medicine and health, resilience studies, child psychology, political science, legal studies, social work, children's rights and ethics, science, and general sustainability research. As a research field, we are indeed contributing to international multidisciplinary research efforts.

In this chapter, we report on our analysis of ECEfS research outputs as measured through published journal papers, special journal issues, and research-based books in ECEfS. We then consider the country of origin for these outputs and elaborate on the key themes emerging from this now quite extensive research collection.

4.2 Measuring ECEfS Research Outputs

Firstly, to seek empirical data on the growth of ECEfS research—beyond our own anecdotal evidence—a diligent process was undertaken, involving four steps. Initially, in March 2023, we employed the Scopus to search for published research articles. Scopus is an abstract and citation database of peer-reviewed literature with tools to track, analyse and visualise research. The selected search words were preschool, early childhood education, kindergarten, and sustainability OR sustainable development. Both these latter descriptors were used, but we note the concept 'sustainability' is not commonly applied in some countries, for example, the USA, Japan, and parts of Europe. We intentionally excluded environmental education (EE) as a search term because our prime focus here is education for sustainability/sustainable development in ECE (see Fig. 4.1), although EE does appear in the titles of some of our selected literature. The screening process was to review and synthesise relevant literature to better understand the field of ECEfS and how it has developed over time, as well as to scrutinise trends and themes. We are aware that the design choices we made provide only a broad understanding of ECEfS and both encourage and challenge other researchers to undertake deeper analyses such as systematic literature reviews.

In the first search, we set time frame parameters for literature published between 2017 and 2023. We note, though, that we have only been able to include publications in English and recognise this as a limitation. This search generated 1227 document results. Then, articles not specifically focused on preschool, early childhood education, kindergarten and sustainability or relevant to ECE were removed, reducing the count to 532-document results. Figure 4.2 was generated using the program

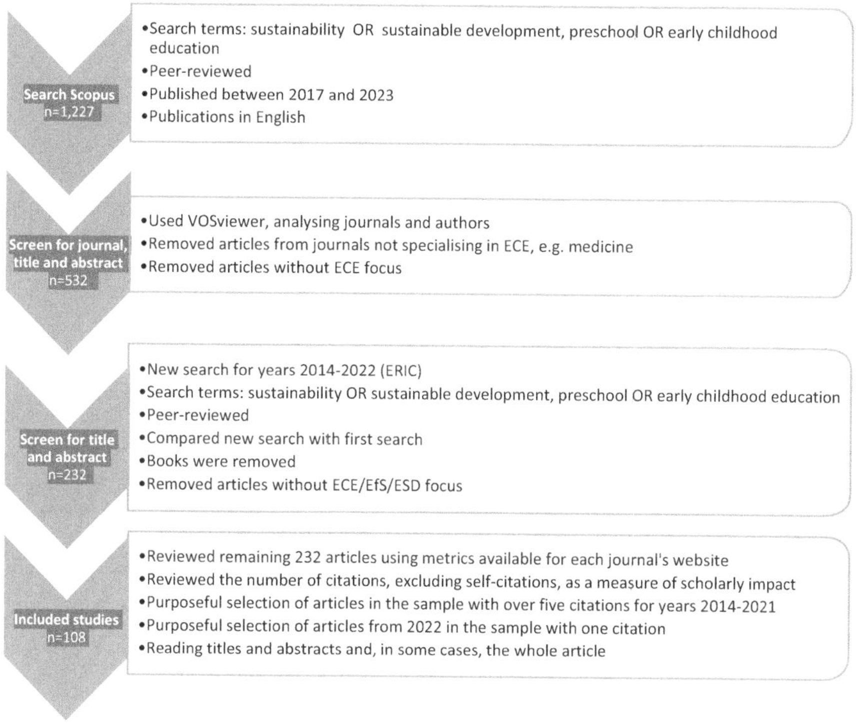

Fig. 4.1 Flow diagram of screening process

VOSviewer to analyse these document results. This figure identifies that the top three journals publishing the largest number of ECEfS articles were in descending order: Sustainability, Environmental Education Research and the International Journal of Early Childhood.

In a second step, we again conducted a search using VOSviewer to locate the most cited authors in the 532-document sample (see Fig. 4.3). This showed that Julie Davis (Australia) and Ingrid Pramling Samuelsson (Sweden) were the most cited researchers from the ECEfS field.

In a third step, articles from journals specialising, for example, in medicine/health were removed from the selection as not primarily ECE focused, though they do confirm the multidisciplinary uptake of ECEfS research. We note that the years 2014–2016 were a period that included several systematic reviews of ECEfS (Green, 2015; Hedefalk et al., 2015; Somerville & Williams, 2015). Next, a new search of the years 2014–2022 was conducted in March 2023 and then compared with the original sample of 532 articles. The starting date of 2014 was chosen to coincide with the first chronology published in 2014 (Davis & Elliott, 2014) and extended upon as outlined in Chap. 2. From this new search step, we gleaned a total of 232 journal articles and 8 research-focused books (see Table 4.6 for these books).

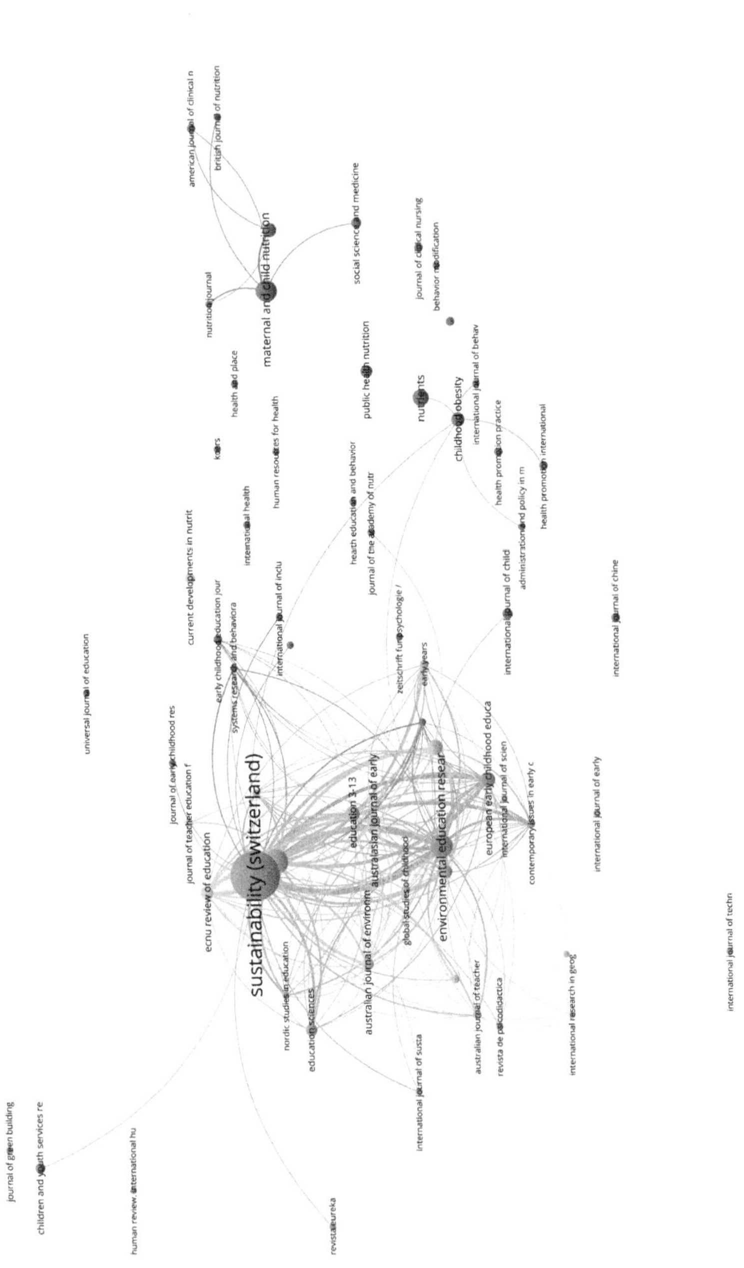

Fig. 4.2 Publishing journals from the 532-document results sample

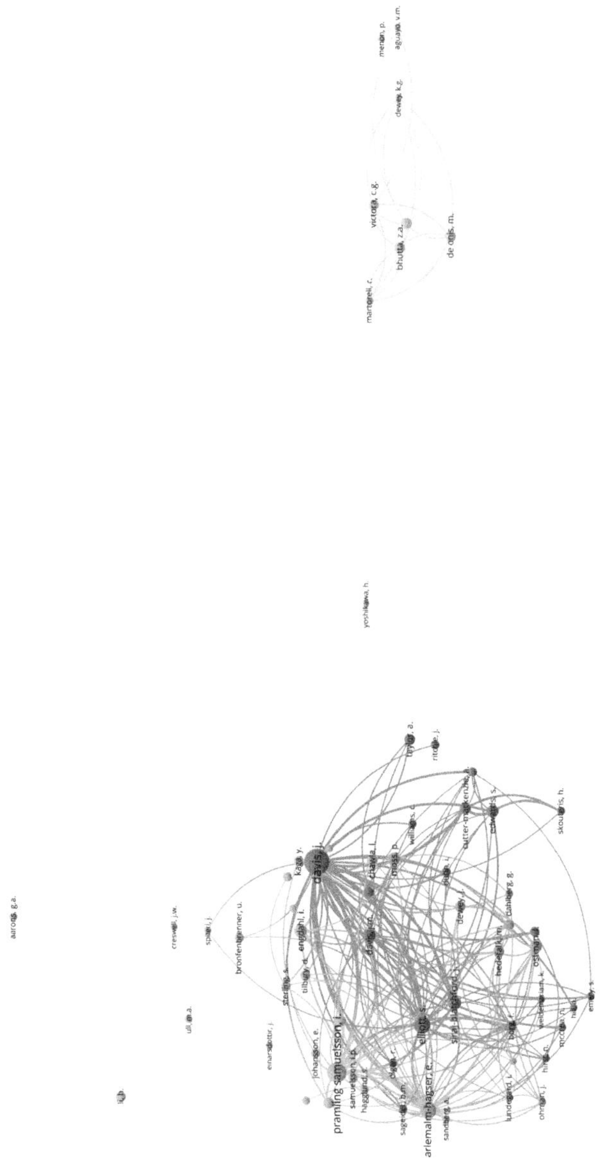

Fig. 4.3 Most cited authors in the 532-document results sample

Table 4.1 Articles 2014–2021 with over five citations and articles in 2022 with one citation

Year	2014	2015	2016	2017	2018	2019	2020	2021	2022
Number	9	8	8	16	11	18	12	15	11

In the fourth and final step, we reviewed each of the 232 articles using metrics available on journal websites, for example, the number of citations, excluding self-citations, as a measure of scholarly impact. We then purposefully selected articles between 2014 and 2022 that had over five citations for the years 2014–2021, and one citation for 2022. This latter figure was chosen because 2022 was too recent to have been cited more extensively (see Table 4.1).

From this final selection round, the number of articles was 108. Thus, Table 4.1 illustrates the ongoing development of the ECEfS research field with the number of articles and citations growing yearly, peaking in 2019. We anticipate that citations will further increase as these publications gain greater reader/viewership.

4.2.1 *Special Journal Issues on ECEfS*

To the best of our knowledge, based on our familiarity with ECEfS research, there have been seven special issues focused on ECE and sustainability published between 2014 and 2022 (see Table 4.2). The topics canvassed reveal ever-widening content that includes, for example, climate change and children's health and well-being, early childhood teacher education, early childhood science education for sustainability, and the SDGs applied to ECE.

These special issues reveal a range of journals, which led us to investigate further which journals published the most ECEfS articles. The seven journals that most frequently published ECEfS research articles are listed in Table 4.3. Of these, the *International Journal of Early Childhood*, *Sustainability*, and *Environmental Education Research* offered the largest article numbers. We have included journal metrics—the 5-year Impact Factor (average number of times articles from the journal have been cited in the past five years) and journal rankings. These metrics tell us that ECEfS research papers are being published in journals with sound reputations, though a future aim could be for researchers to select those with higher rankings as the field advances. In addition, there were at least 38 journals globally each publishing 1–5 articles during the specified timeframe from the total database of 108 document results.

Table 4.2 Special journal issues on ECEfS from 2014 to 2022

Editors and journal	Content	Context
Currie, J., & Deschênes, O. (2016). Children and climate change: Introducing the issue. *The Future of Children, 26*(1), 3–9. https://doi.org/10.1353/foc.2016.0000	Special Issue: *Health impacts for children*	International
Ärlemalm-Hagsér, E., & Elliott, S. (2017). Special Issue. Contemporary research on early childhood education for sustainability. *International Journal of Early Childhood, 49*(3), 267–272. https://doi.org/10.1007/s13158-017–0207-3	Special Issue: *Education for sustainability*	International
Ärlemalm-Hagsér, E., & Hedefalk, M. (2018). Editorial. Förskolan och utbildning för hållbarhet—Nordisk forskning inom fältet förskola och hållbarhet. [Special issue editorial: Preschool and education for sustainability—Nordic research in the field of preschool and sustainability]. *Utbildning & Demokrati, 27*(2), 7–14. https://doi.org/10.48059/uod.v27i2.1099	Special Issue: *Sustainability in the Nordic countries*	Nordic countries
Sageidet, B. M. et al. (2019). Editorial. *NorDiNa* - Nordic Studies in Science Education special issue sustainability and science education in kindergarten, *15*(4), 338–340. https://doi.org/10.5617/nordina.7421	Special Issue: *Sustainability and science education*	International
Carr, V., Elliott, S., & Ärlemalm-Hagsér, E. (2021). Editorial. Special Issue: *International Journal of Early Childhood Environmental Education*—Sustainable Development Goals and Early Childhood Environmental Education: Awareness, understanding, and transformation, *9*(1), 5–8. https://naturalstart.org/research/ijecee/volume-9-number-1	Special Issue: *UNESCO SDGs*	International
Engdahl, I., & Furu, A-C. (2022). Early childhood education for sustainability. *International Journal of Early Childhood, 54*(1). https://link.springer.com/journal/13158/volumes-and-issues/54-1	Special Issue: *Education for sustainability*	International
Evans, N., Harwood, D., Furu, A.-C., & Alici, Ş. (2022). Challenges in reorienting early childhood education for sustainability in initial teacher education: Transforming, creating and expanding. *Australian Journal of Teacher Education, 47*(5). https://doi.org/10.14221/ajte.2022v47n5.1	Special Issue: *Teacher education, initial teacher education*	International

4.2.2 Research Outputs by Country of Origin

The origins of the 108 articles in the sample illustrated that international author collaborations were common, with authors from countries such as Sweden, Australia,

Table 4.3 Seven journals from the sample in which ECEfS research is most frequently published

Journal title	No. of articles on ECEfS	Journal metrics (5-yr IF) (2022)	Journal rank
International Journal of Early Childhood	24	4.4	Q2 (Citescore)
Sustainability	18	4.0	Q2 (JCR)
Environmental Education Research	8	3.9	Q1 (Citescore)
Education 3–13	4	1.0	Q2 (Citescore)
International Journal of Early Childhood Environmental Education	4	1.4	Not yet ranked
Australasian Journal of Early Childhood	3	1.1	Q2 (Scimago)
European Early Childhood Education Research Journal	3	2.1	Q2 (Citescore)

Türkiye, and the USA most strongly represented (see Table 4.4). We can only speculate as to why a limited range of author nationalities was evident. Perhaps journal access and language barriers, along with limited support for ECEfS research on the ground, or lack of ECEfS research 'champions' amidst national priorities in ECE, were factors. We argue there is scope to examine publications from majority worlds and those published in languages other than English. For example, anecdotally we know of relevant publications in Japan and Korea, but these countries were not represented here because the publications searched were only in English.

4.3 Research Themes

Among the 108 articles selected by reading the titles and abstracts—and in some cases reading the whole article—the most frequent themes were determined. These were: children's participation, agency and action competence; children's learning and understandings; teachers' understandings and professional learning EfS; ECEfS pedagogies; sustainability content areas; sustainability in national curriculum policy; early childhood teacher education; and methodological and theoretical perspectives in ECEfS research (see Table 4.5). We discuss and elaborate these themes following the table. Less frequent minor themes including leadership, citizenship, and country updates regarding curriculum policy and policy vacuums were also evident.

Table 4.4 Author's country of origin for 108 articles from 2014 to 2022

Author's country of origin	No. of peer-reviewed articles
International collaboration	20
Sweden	21
Australia	15
Türkiye	10
USA	10
Norway	7
Malta	3
New Zealand	3
UK	3
Canada	2
Denmark	2
Finland	2
Indonesia	2
Malaysia	2
Spain	2
Belgium	1
Greece	1
Iceland	1
Singapore	1
South Korea	1

4.3.1 Theme 1: Children's Participation, Agency, and Action Competence

Over 15 years ago, Davis (2009) reported that there were very few articles on ECEfS in general (fewer than 5% in a selection of journals in which ECE researchers regularly published) and even fewer articles focused children's active learning and participation in sustainability. Since then, EfS research prioritising children as active participants has developed apace (Ärlemalm-Hagsér, 2014; Ärlemalm-Hagsér & Davis, 2014; Caiman & Lundegård, 2014; Engdahl, 2015; Hedefalk et al., 2015). However, there has been little attention paid to how children's agency is pedagogically supported by teachers in the research literature (Borg & Pramling Samuelsson, 2022; Mackey & De Vocht, 2016). In the context of the UNCRC (UNCRC, 1989), the impacts of climate change on children and contemporary ECE theorising about children's capacities and agencies, there is scope for many more studies focusing on children's agency.

Table 4.5 Most frequent themes within the selected 108 articles from 2014 to 2022

Year of publication	Research themes						
2014	Children's participation, agency and action competence	Teacher understandings	Content	Children's learning and understandings	Pedagogy	Country updates	
2015	Methodological and theoretical perspectives	Teacher understandings	Content	Children's learning and understandings	Curriculum policy		
2016	Children's agency	Teacher understandings	Content	Teacher education	Pedagogy		
2017	Teacher education	Teacher understandings	Content	Pedagogy	Children's learning	Curriculum policy	Methodological and theoretical perspectives
2018	Methodological and theoretical perspectives	Teacher understandings	Content	Pedagogy	Curriculum policy		
2019	Teacher education	Teacher understandings	Content	Pedagogy	Children's learning and understandings	Citizenship	
2020	Teacher education	Children's learning and understandings	Content	Pedagogy	Leadership	Teacher understandings	Methodological and theoretical perspectives
2021	Teacher education	Children's understandings	Pedagogy	Methodological and theoretical perspectives	Citizenship	Curriculum policy	Methodological and theoretical perspectives
2022	Teacher education	Children's agency	Content	Pedagogy	Teacher understandings		

4.3.2 Theme 2: Children's Learning and Understandings

Aspects of this theme were diverse: children as eco-citizens (Heggen et al., 2019), children's awareness of their ecological footprint and environmental attitudes (Simsar, 2021), the environmental impacts of various transport modes (Borg et al., 2019), and children's environmental identity (Green, 2017). Several researchers explored children's ideas about the environment (Madden & Liang, 2017; Melis et al., 2020; Spiteri et al., 2022), or specifically, how children understand environmental issues (Spiteri, 2021a, 2021b). In some research studies, children's understandings and learning with different tools (Zurek et al., 2014), as well as collaboration and inquiry through projects (Feriver, 2022; Hirst, 2019) and places, were emphasised as important to their sustainability learning and skill development. For example, nature play was seen as constructing curiosity, creative thinking, and resilience in relation to sustainability (Ernst & Burcak, 2019). Other aspects included children's sustainability learning within the home setting and its relationship with preschool-based learning about sustainable societies (Borg et al., 2017). There appears no shortage of avenues to investigate where, when, and how children are learning and understanding about living sustainably, illustrating the significant potential for ECE as a foundation for EfS.

4.3.3 Theme 3: Teachers' Understandings and Professional Learning

Studies investigating teachers' understandings around ECEfS demonstrated that they were unsure about what EfS meant and how it could be realised in ECE settings (Inoue et al., 2016; Samuelsson Pramling & Park, 2017). It was reported they struggled to handle the complexity and ambiguity of sustainability issues (Ärlemalm-Hagsér & Sundberg, 2016; Hedefalk et al., 2015) and often appeared to interpret EfS in ECE differently (Hedefalk et al., 2015; Hill, et al., 2014; Lasen et al., 2017; Višnjic-Jevtic et al., 2022). Studies about how teachers pursued their identity were seen as key for understanding their critical role in EfS (Almeida et al., 2018).

Teachers' beliefs and views were also studied from different perspectives: being outside (Norðdahl & Jóhannesson, 2016); their nature relatedness in EfS (Sahin & Alici, 2019); EE and environmental awareness (Türkoğlu, 2019); teachers' views of socio-environmental problems (Rodríguez-Marín et al., 2020); multicultural sustainability (Huda et al., 2021); and inclusive education (You et al., 2019). It would appear understanding the beliefs and views they bring to teaching EfS with young children is diverse and pivotal for how early childhood settings embed sustainability. Teachers also demonstrated a tendency to describe EfS as something they do in their everyday practices, just routine (Engdahl et al., 2021). However, anecdotally, we recognise teachers are struggling to find new ways to meet the demands of governing policy documents, with new sustainability goals connected to EfS becoming more evident in national curricula.

In the 108 collated studies, EC teachers also articulated the necessity for professional development about ECEfS (Bascopé et al., 2019; Dyment, et al., 2014; Feriver et al., 2016) and how to create transformative change in their everyday practices (Engdahl et al., 2021; Wakefield et al., 2022). For example, Borg and Gericke's (2021) study illustrated that after teachers participated in an EfS-school development project, they began to integrate the social dimension of sustainability into their teaching and associate local sustainability challenges with those faced internationally.

4.3.4 Theme 4: ECEfS Pedagogies

In several articles, different pedagogies and programs were discussed and evaluated demonstrating both how to handle EfS in practice and as a theoretical discussion. For example, Willockx and Dom (2022) described EfS in practice using the UNESCO SDGs as a theoretical framework for implementing sustainability in ECE. Boyd (2019) utilised place-based learning theory in local contexts to promote children as agents of change. She also evaluated an intergenerational sustainability project (Boyd, 2020) where children learnt arts, crafts, and traditional skills from elders. Other studies focused on creative processes (Caiman et al., 2022), collaborative exploration (Ødegaard, 2021), and the importance of conversations both planned and incidental for promoting EfS (Bautista et al., 2018).

Recognised programs, such as the eco-school program, were discussed and evaluated (Kahriman-Pamuk & Olgan, 2018; Korkmaz & Yildiz, 2017). In other studies, pedagogical explorations included wild pedagogies (Jickling et al., 2018), garden education (Lee et al., 2019), and nature-based approaches (Carter, 2016; Inoue et al., 2019), as children's nature play is often seen as foundational in ECEfS (Ernst et al., 2021). There were also articles that discussed the need for theoretical perspectives beyond 'weak' EE/EfS stewardship ideas (Taylor, 2017) and the importance of theorising ECE in the Anthropocene (Wolff et al., 2020).

4.3.5 Theme 5: Sustainability Content in ECEfS

The sample studies demonstrated that the ECEfS field has expanded across many content areas in ECE, including the social, economic, and environmental dimensions of sustainability.

Content areas related to social sustainability included children's belonging in peer communities (Johansson & Rosell, 2021); care (Wals, 2017; Weldemariam et al., 2022); children's views on equality, work, money, and access to food (Hammond et al., 2015); fostering eco-cultural literacies for social, cultural, and ecological justice (Ritchie, 2017); food reciprocity and sustainability (Ritchie, 2015); well-being (Edwards et al., 2016); food choices and sustainability (Boyd, 2015); and early literacy education (Schmidt, 2017). In addition, Nxumalo (2018) employed language and stories to support EE in a preschool classroom, and Tzima et al. (2020) explored storytelling and mobile technology in relation to cultural heritage in ECE. Content areas in the sample connected to economic dimension of sustainability included children's views of economic (in)equality and sustainability (Borg, 2017); preschool teachers and preschools as political-economic actors (Ärlemalm-Hagsér et al., 2018); and migrant children and social sustainability (Boldermo & Eriksen Ødegaard, 2019).

Unsurprisingly, the largest number of articles about ECEfS content areas focused on the environmental dimension and related nature/environmental issues. These included weather (Rooney, 2018), bees (Weldemariam, 2020), human and nonhuman belonging (Sadownik & Gabi, 2021), and the importance of animals in ECE settings (Born, 2018). Nature experiences and gardening content were well described as part of ECE practices (Ärlemalm-Hagsér & Sundberg, 2016; Bergan, et al., 2021; Jorgensen, 2016), but Elliott and Young (2016) caution the need for critical awareness as 'nature by default' does not develop children's understandings of sustainability.

One content area described in several articles was the now common habits of recycling or sorting waste (Buil et al., 2019; Jørgensen, Madsen & Læssøe, 2018; Rahmayanti et al., 2020; Pamuk & Kahriman-Pamuk, 2019). Also, different aspects of science and sustainability were examined, for example, technology (Willis et al., 2014), multidimensional preschool science (Sundberg et al., 2019), and water projects (Ampartzaki et al., 2021) to deepen children's knowledge.

Finally, a few articles described art and sustainability. For example, Chapman and O'Gorman (2022) discussed how to transform an ECE learning environment through the arts in response to the UN SDGs, while Akyol et al., (2018) described how drama was used with teachers. In addition, music and literacy as content areas can promote ECEfS practices (Powell & Somerville, 2020).

4.3.6 Theme 6: Sustainability in National Curriculum Policy

A few articles were framed as curriculum policy analyses canvassing issues about national curricula and sustainability (Ärlemalm-Hagsér & Davis, 2014; Moore et al., 2018; Pollock et al., 2017). Specifically, Weldemariam et al. (2017) compared a number of different national frameworks and one article described curriculum development in the context one early childhood centre (MacDonald, 2015). Anecdotally, however, we recognise that there is much to do in terms of a critical mass of countries having sustainability embedded into their national curriculum documents. We are still at the beginning of this policy agenda.

4.3.7 Theme 7: Early Childhood Teacher Education

Among the articles sampled, a small number of initial teacher education studies were evident. The topics of these articles were: preschool teacher students' EfS meaning-making, knowledge, and attitudes (Ärlemalm-Hagsér & Larsson, 2019; Mahat et al., 2019); exploring student teachers' workplace experiences (practicum) with respect to EfS understandings (Ärlemalm-Hagsér, 2017); and problematising didactical dilemmas when planning for sustainability teaching (Hedefalk et al., 2021). More broadly, Weldemariam (2022) discussed the ontological and epistemological (from a post-humanist stance) perspectives of EfS in relation to broadening student teachers' understandings of sustainability discourses. Furu et al. (2021) suggested multimodal storytelling to promote understandings about sustainability, while Skarstein and Skarstein (2020) examined student teachers' species knowledge and identification skills. While there is a small amount of research occurring around this theme, we know anecdotally that EC teacher education for sustainability is still a weak spot in practice and must be addressed.

4.3.8 Theme 8: Methodological and Theoretical Perspectives in ECEfS Research

From being almost non-existent in the earliest days in the evolution of ECEfS (Davis & Elliott, 2003), discussion of methodological and theoretical perspectives in ECEfS is now being extended through literature reviews and various targeted theoretical papers, for example, analyses of the ECEfS research literature (Ardoin & Bowers, 2020; Yıldız et al., 2021), and theorising about post-humanist perspectives (Somerville & Williams, 2015; Taylor, 2017). An analysis of ECEfS methodologies and methods in research (Green, 2015), feminist and post-colonial ethics of care (Raivio et al., 2022), and post-humanist perspectives on cities as key sites for reimagining and reconfiguring human-nature encounters (Duhn et al., 2017) offer deeper insights into how and why ECEfS might be enacted. We anticipate that areas of future theorising and critique of ECEfS will see a deepening of First Nations and non-Western perspectives on ECEfS.

4.4 Additional Research Impact Measures

The publication data analysis process described above only recorded research articles published in peer-reviewed academic journals. However, we know from our extensive knowledge of the field that this significantly under-represents ECEfS research outputs. For example, while there has been a plethora of excellent publications with a focus on practitioners, we explicitly focus on research outputs here.

4.4.1 Research and Research-Informed Books in ECEfS

In Table 4.6, we add to the pool of research outputs identified so far in this chapter, with a set of research and research-informed books published (in English) between 2014 and 2022. We know from alerts through websites (such as Researchgate, Google Scholar and Academia.org) that many chapters in these books are regularly cited, but unable to be captured through programs such as VOSviewer. We note there have been translations of two of these texts, another measure of international impact.

Table 4.6 Research and research-informed books on ECE and sustainability (2014 to 2022)

Title of research books	Country of origin
Cutter-Mackenzie, A., Edwards, S., Moore, D., & Boyd, W. (2014). *Young children's play and environmental education in early childhood education* (2014.). Springer International Publishing. https://doi.org/10.1007/978-3-319–03740-0	Australia
Davis. J. & Elliott, S. (2014). *Research in early childhood education for sustainability: International perspectives and provocations.* Routledge	International; translated into Korean (2014)
Davis. J. (2015). *Young children and the environment: Early education for sustainability* (2nd ed.). Cambridge University Press	International; translated into Chinese (2017)
Siraj-Blatchford, J., Mogharreban, C., & Park, E. (Eds.) (2016). *International research on education for sustainable development in early childhood.* Springer International Publishing. https://doi.org/10.1007/978-3-319–42208-4	International
Boyd, D., Hirst, N., & Siraj Blatchford, J. (2018). *Understanding sustainability in early childhood education: Case studies and approaches from across the UK.* Routledge	United Kingdom
Huggins, V., & Evans, D. (2018). *Early childhood care and education for sustainability.* Routledge	International
Cutter-Mackenzie-Knowles, A., Malone, K., & Barratt Hacking, E. (2020). *Research handbook on childhoodnature: Assemblages of childhood and nature research.* Springer International Publishing. https://doi.org/10.1007/978-3-319–51949-4	International
Elliott, S., Ärlemalm-Hagsér, E., & Davis, J. (2020). *Researching early childhood education for sustainability: Challenging assumptions and orthodoxies.* Routledge	International

4.4.2 Further Research Impact Measures

We do not have the means to capture all possible research outputs in ECEfS. Below is a short list of just some of these additional outputs:

- ECEfS authors have been invited to write bibliographies, e.g. Oxford University Press (Ärlemalm-Hagsér & Elliott, 2021; Davis, 2022; Ritchie, 2021) and encyclopaedia entries (Elliott & Ärlemalm-Hagsér, 2019). Such invitations are a measure of our profiles as ECEfS leaders, and recognition of ECEfS as having reached a level of importance to be worthy of inclusion in such catalogues that are not ECEfS specific.
- ECEfS researchers are consistently and increasingly invited as keynote speakers at local, national, and international conferences and symposia, for example in Italy, Japan, Korea, New Zealand, Singapore, and the USA. We produce videos, webinars, and blogs and regularly translate our research into publications for the ECE profession and the general public.
- There has been an upswing of original research through Masters and doctoral theses, again a sign of growing research maturity and strength. Examples include:

  Alici, Ş. (2018). *Investing the impact of professional development on Turkish early childhood teachers' professional growth about education for sustainable development through critical media literacy.* PhD. Middle East Technical University, Ankara, Türkiye.

  Christensen, M. (2021). *Children's understandings of sustainability related topics and issues: A phenomenographic investigation seen through drawings and interviews with 6–8-year-old children.* Professional Doctorate thesis, QUT, Australia.

  Dean, S. (2022). *A phenomenological study of the lived experiences of teachers who implement outdoor/environmental education within a k-12 setting.* Clemson University, USA.

  Hughes, F. (2021). *Problematising sustainability in nature play programs: Pedagogical transformations and ambitious alignments.* (Doctor of Philosophy). University of New England, Australia.

  Kim, S. (2022). *A comparative case study of Korean and Australian early childhood education for sustainability: Exploring teachers' practices, beliefs, and assumptions through a cultural lens.* PhD thesis, QUT, Australia.

  Paujik, Y. (2019) *Young children's understanding of poverty within a framework of sustainability: Action research in a kindergarten.* M. Philosophy thesis, QUT, Australia.

  Weldemariam, K. (2020). *Reconfiguring environmental sustainability in early childhood education: A post-anthropocentric approach.* PhD thesis. University of Gothenburg, Sweden.

Several theses have been reworked into books, book chapters, and peer-reviewed articles. For example, Hughes' thesis has been rewritten as a book (Hughes et al., 2022).

- There has been utilisation of ECEfS research in national and international reports, often without attribution to specifically named authors. While we are unable to account for these publications in this summary of ECEfS research outputs, we know they have been impactful for national and international policy development. This includes contributions to:

  UNESCO (2014) *Shaping the future we want,* Global Monitoring and Evaluation Report of *the UN Decade of Education for Sustainable Development (2005–2014);*
  United Nations Climate Change conferences (CoP26, 2021 and CoP27, 2022);
  European Commission initiatives on *Learning for the Green Transition and Sustainable Development* (2022); and
  Updates to national curricula such as the Australian Learning Frameworks (AGDE, 2022).

4.5 Conclusion

In this chapter, we have focused our attention on the role of research in ECEfS. As seen above, the themes of children's participation, agency and action competence, children's learning and understandings, teacher's understandings and professional learning in sustainability, ECEfS pedagogies, sustainability content areas, sustainability in curriculum policy, early childhood teacher education, and methodological and theoretical perspectives in ECEfS research have been investigated and discussed from the 2014–2022 article sample. Other less prominent themes included ECEfS leadership, ECEfS and citizenship, and curriculum policy country updates. These latter themes point to potential new foci for forthcoming research. Our basic premise is that research is an important and powerful enabler for the continued growth of the field. Fortunately, we are not alone in this thinking as demonstrated by the ever-increasing pool of studies informing efforts this expanding field across peer-reviewed papers and published books about ECEfS research. The power of research is that it moves people away from their personal feelings of what is good for/in ECEfS and ties it to the rigours of critical thinking, evidence making and peer review. Further, it links the field to major national and international reports and enables access to multidisciplinary inputs, while carving out the uniqueness of ECEfS that derives from its historical precedents and contemporary, ever-changing contexts.

Appendix: List of the 108 Selected Research Publications that Chart ECEfS Development (2014–2022)

Year	Publication	Focus	Origin
2014			
1	Agut, M. P. M., Ull, M. A., & Minguet, P. A. (2014). Education for sustainable development in early childhood education in Spain. Evolution, trends and proposals. *European Early Childhood Education Research Journal*, *22*(2), 213–228. https://doi.org/10.1080/1350293X.2013.783299	Country updates	Spain
2	Ärlemalm-Hagsér, E. (2014). Participation as "taking part in": Education for sustainability in Swedish preschools. *Global Studies of Childhood*, *4*(2), 101–114. https://doi.org/10.2304/gsch.2014.4.2.101	Children's participation	Sweden
3	Ärlemalm-Hagsér, E., & Davis, J. (2014). Examining the rhetoric: A comparison of how sustainability and young children's participation and agency are framed in Australian and Swedish early childhood education curricula, *Contemporary Issues in Early Childhood*, *15*(3), 231–244. https://doi.org/10.2304/ciec.2014.15.3.231	Curriculum policy and children's participation	Sweden and Australia
4	Caiman, C., & Lundegård, I. (2014). Pre-school children's agency in learning for sustainable development. *Environmental Education Research*, *20*(4), 437–459. https://doi.org/10.1080/13504622.2013.812722	Children's agency	Sweden
5	Dyment, J. E., Davis, J. M., Nailon, D., Emery, S., Getenet, S., McCrea, N., & Hill, A. (2014). The impact of professional development on early childhood educators' confidence, understanding and knowledge of education for sustainability. *Environmental Education Research*, *20*(5), 660–679. https://doi.org/10.1080/13504622.2013.833591	Early childhood educators Confidence, understanding and knowledge of education	Australia

(continued)

(continued)

Year	Publication	Focus	Origin
6	Hedefalk, M., Almqvist, J., & Lidar, M. (2014). Teaching for action competence. *SAGE Open*, *4*(3). https://doi.org/10.1177/2158244014543785	Children's action competence	Sweden
7	Hill, A., Nailon, D., Getenet, S., McCrea, N., Emery, S., Dyment, J., & Davis, J. M. (2014). Exploring how adults who work with young children conceptualise sustainability and describe their practice initiatives. *Australasian Journal of Early Childhood*, *39*(3), 14–22. https://doi.org/10.1177/183693911403900303	Early childhood educators Pedagogy	Australia
8	Willis, J., Weiser, B., & Kirkwood, D. (2014). Bridging the Gap: Meeting the needs of early childhood students by integrating technology and environmental education. *International Journal of Early Childhood Environmental Education*, *2*(1), 140–155	Content Technology and environmental education with children	USA
9	Zurek, A., Torquati, J., & Acar, I. (2014). Scaffolding as a tool for environmental education in early childhood. *International Journal of Early Childhood Environmental Education*, *2*(1), 27–52	Scaffolding children in environment education Children's learning	USA
2015			
1	Boyd, W., (2015). The tensions between food choices and sustainability practices in early childhood education centres. *Australasian Journal of Early Childhood*, *40*(4), 58–65. https://doi.org/10.1177/183693911504000408	Content Food choices and sustainability practices in ECE centres	Australia
2	Engdahl, I. (2015). Early childhood education for sustainability: The OMEP world project. *International Journal of Early Childhood*, *47*(3), 347–366. https://doi.org/10.1007/s13158-015-0149-6	Children's participation The OMEP world project	International

(continued)

(continued)

Year	Publication	Focus	Origin
3	Green, C. (2015). Toward young children as active researchers: A critical review of the methodologies and methods in early childhood environmental education. *Journal of Environmental Education*, *46*(4), 207–229, https://doi.org/10.1080/00958964.2015.1050345	Methodological and theoretical perspectives Review of ECEE research Methodologies/methods	USA
4	Hammond, L. L., Hesterman, S., & Knaus, M. (2015). What's in your refrigerator? Children's views on equality, work, money and access to food. *International Journal of Early Childhood*, *47*(3), 367–384. https://doi.org/10.1007/s13158-015-0150-0	Content Children's views on equality, work, money, food, access, social and economic sustainability	Australia
5	Hedefalk, M., Almqvist, J., & Östman, L. (2015). Education for sustainable development in early childhood education: A review of the research literature. *Environmental Education Research*, *21*(7), 975–990. https://doi.org/10.1080/13504622.2014.971716	Teacher understandings Review of research literature	Sweden
6	MacDonald, M. (2015). Early childhood education and sustainability: A living curriculum. *Childhood Education*, *91*(5), 332–341. https://doi.org/10.1080/00094056.2015.1090845	Curriculum policy	Canada
7	Ritchie, J. (2015). Food reciprocity and sustainability in early childhood care and education in Aotearoa New Zealand. *Australian Journal of Environmental Education*, *31*(1), 74–85. https://doi.org/10.1017/aee.2014.46	Content Food reciprocity and sustainability	New Zealand
8	Somerville, M. & Williams, C. (2015). Sustainability education in early childhood: An updated review of research in the field. *Contemporary Issues in Early Childhood*, *16*(2), 102–117. https://doi.org/10.1177/1463949115585658	Literature overview Theoretical and methodological developments informing research in ECE sustainability education	International

(continued)

(continued)

Year	Publication	Focus	Origin
2016			
1	Ärlemalm-Hagsér, E., & Sundberg, B. (2016). Nature experiences and recycling—A quantitative study on education for sustainable development in Swedish preschools. *NorDina, 12*(2), 140–156. https://doi.org/10.5617/nordina.1107	Content areas Quantitative study on education for sustainable development	Sweden
2	Carter, D. (2016). A nature-based social-emotional approach to supporting young children's holistic development in classrooms with and without walls: The Social-Emotional and Environmental Education Development (SEED) Framework. *International Journal of Early Childhood Environmental Education, 4*(1), 9–24	Pedagogy Nature Environmental education	USA
3	Edwards, S., Skouteris, H., Cutter-Mackenzie, A., Rutherford, L., O'Conner, M., Mantilla, A., Morris, H., & Elliott, S. (2016). Young children learning about well-being and environmental education in the early years: A funds of knowledge approach. *Early Years, 36*(1), 33–50. https://doi.org/10.1080/09575146.2015.1064099	Content Randomised trial examining interconnections between sustainability, well-being and popular culture	Australia
4	Elliott, S. & Young, T. (2016). Nature by default in early childhood education for sustainability. *Australian Journal of Environmental Education, 32*(1), 57–64. https://doi.org/10.1017/aee.2015.44	Content Nature and sustainability	Australia
5	Feriver, Ş., Teksöz, G. (Tuncer), Olgan, R., & Reid, A. (2016). Training early childhood teachers for sustainability: Towards a "learning experience of a different kind". *Environmental Education Research, 22*(5), 717–746. https://doi.org/10.1080/13504622.2015.1027883	Training early childhood teachers for sustainability	Türkiye

(continued)

(continued)

Year	Publication	Focus	Origin
6	Jorgensen, K.-A. (2016). Bringing the jellyfish home: Environmental consciousness and "sense of wonder" in young children's encounters with natural landscapes and places. *Environmental Education Research*, *22*(8), 1139–1157. https://doi.org/10.1080/13504622.2015.1068277	Content Nature-kindergarten ethnography Environmental education	Norway
7	Mackey, G., & De Vocht, L. (2016). Teachers explore how to support young children's agency for social justice. *International Journal of Early Childhood*, *48*(3), 353–367. https://doi.org/10.1007/s13158-016-0175-z	Children's agency	New Zealand
8	Norðdahl, K., & Jóhannesson, I. Á. (2016). "Let's go outside": Icelandic teachers' views of using the outdoors. *Education 3–13*, *44*(4), 391–406. https://doi.org/10.1080/03004279.2014.961946	Teacher understandings of the role of outdoor environments in children's learning	Iceland
2017			
1	Ärlemalm-Hagsér, E. (2017). Student teachers' workplace-based learning in Sweden on early childhood education for sustainability: Experiences in practice settings. *International Journal of Early Childhood*. *49*(3), 411–427. https://doi.org/10.1007/s13158-017-0201-9	Teacher education and sustainability	Sweden
2	Borg, F. (2017). Economic (in)equality and sustainability: Preschool children's views of the economic situation of other children in the world. *Early Child Development and Care*, first online 1–15. https://doi.org/10.1080/03004430.2017.1372758	Content Preschool children's views of the economic situation of children in the world	Sweden
3	Borg, F., Winberg, M., & Vinterek, M. (2017). Children's learning for a sustainable Society: Influences from home and preschool. *Education Inquiry*, *8*(2), 151–172. https://doi.org/10.1080/20004508.2017.1290915	Children's learning for a sustainable Society	Sweden

(continued)

(continued)

Year	Publication	Focus	Origin
4	Duhn, I., Malone, K., & Tesar, M. (2017). Troubling the intersections of urban/nature/childhood in environmental education. *Environmental Education Research, 23*(10), 1357–1368. https://doi.org/10.1080/13504622.2017.1390884	Methodological and theoretical perspectives Post-humanism urban/nature/childhood	Australia/ New Zealand
5	Green, C. (2017). Children environmental identity development in an Alaska native rural context. *International Journal of Early Childhood, 49*(3), 303–319. https://doi.org/10.1007/s13158-017-0204-6	Children's learning Children environmental identity development in an Alaskan context	USA
6	Inoue, M., O'Gorman, L., Davis, J., & Ji, O. (2017). An international comparison of early childhood educators' understandings and practices in education for sustainability in Japan, Australia and Korea. *International Journal of Early Childhood, 49*(3), 353–373. https://doi.org/10.1007/s13158-017-0205-5	Teacher understandings International comparison of early childhood educators' understandings and practices	Japan, Australia, and Korea
7	Korkmaz, A., & Guler Yildiz, T. (2017). Assessing preschools using the Eco-Schools program in terms of educating for sustainable development in early childhood education. *European Early Childhood Education Research Journa*l, *25*(4), 595–611. https://doi.org/10.1080/1350293X.2017.1331074	Pedagogy Eco-schools program	Türkiye
8	Lasen, M., Skamp, K., & Simoncini, K. (2017). Teacher perceptions and self-reported practices of education for sustainability in the early years of primary school: An Australian case study. *International Journal of Early Childhood, 49*(3), 391–410. https://doi.org/10.1007/s13158-017-0200-x	Teacher perceptions and self-reported practices of education for sustainability	Australia
9	Madden, L., & Liang, J. (2017). Young children's ideas about environment: Perspectives from three early childhood educational settings. *Environmental Education Research, 23*(8),1055–1071. https://doi.org/10.1080/13504622.2016.1236185	Children's understandings of environmental sustainability education	USA

(continued)

(continued)

Year	Publication	Focus	Origin
10	Ritchie, J. (2017). Fostering eco-cultural literacies for social, cultural and ecological justice: A perspective from Aotearoa (New Zealand). *International Journal of Early Childhood*, *49*(3), 287–302. https://doi.org/10.1007/s13158-017-0198-0	Content Eco-cultural literacies for social, cultural, and ecological justice	New Zealand
11	Samuelsson Pramling I., & Park, E. (2017). How to educate children for sustainable learning and for a sustainable world. *International Journal of Early Childhood*, *49*(3), 273–285. https://doi.org/10.1007/s13158-017-0197-1	Teacher understandings Teaching How to educate children	Sweden and South Korea
12	Schmidt, C. (2017). Thrown Together: Incorporating place and sustainability into early literacy education. *International Journal of Early Childhood*, *49*(2), 165–179. https://doi.org/10.1007/s13158-017-0192-6	Content Early literacy education	Sweden
13	Taylor, A. (2017). Beyond stewardship: Common world pedagogies for the Anthropocene. *Environmental Education Research*, *23*(10), 1448–1461. https://doi.org/10.1080/13504622.2017.1325452	Methodological and theoretical perspectives Pedagogies for the Anthropocene	Australia
14	Pollock, K., Warren, J., & Andersen, P. (2017). Inspiring environmentally responsible preschool children through the implementation of the National Quality Framework: Uncovering what lies between theory and practice. *Australasian Journal of Early Childhood*, *42*(2), 12–19. https://doi.org/10.23965/AJEC.42.2.02	Curriculum policy implementation of the Australian National Quality Framework	Australia
15	Wals, A. E. (2017). Sustainability by default: Co-creating care and relationality through early childhood education. *International Journal of Early Childhood*, *49*(2), 155–164. https://doi.org/10.1007/s13158-017-0193-5	Content Co-creating care and relationality	International

(continued)

(continued)

Year	Publication	Focus	Origin
16	Weldemariam, K., Boyd, D., Hirst, N., Sageidet, B. M., Browder, J. K., Grogan, L., & Hughes, F. (2017). A critical analysis of concepts associated with sustainability in early childhood curriculum frameworks across five national contexts. *International Journal of Early Childhood, 49*(3), 333–351. https://doi.org/10.1007/s13158-017-0202-8	Curriculum frameworks	International
2018			
1	Akyol, T., Kahriman-Pamuk, D., & Elmas, R. (2018). Drama in education for sustainable development: Preservice preschool teachers on stage. *Journal of Education and Learning, 7*(5), 102–115. https://www.ccsenet.org/journal/index.php/jel/article/view/74316	Content Drama in education for sustainability	Türkiye
2	Almeida, S. C., Moore, D., & Barnes, M. (2018). Teacher identities as key to environmental education for sustainability implementation: A study from Australia. *Australian Journal of Environmental Education, 34*(3), 228–243. https://doi.org/10.1017/aee.2018.40	Teacher understandings Teacher identities EEfS Implementation	Australia
3	Ärlemalm-Hagsér, E., Berg, B., & Sandberg, A. (2018). Economic sustainability in Swedish preschools. Preschool teachers and preschools as political-economic actors. *Utbildning & Demokrati, 27*(2), 15–36. https://doi.org/10.48059/uod.v27i2.1100	Content Economic sustainability in Swedish preschools	Sweden
4	Bautista, A., Moreno-Núñez, A., Ng, S.-C., & Bull, R. (2018). Preschool educators' interactions with children about sustainable development: Planned and incidental conversations. *International Journal of Early Childhood, 50*(1), 15–32. https://doi.org/10.1007/s13158-018-0213-0	Pedagogic Preschool educators' interactions with children	Singapore

(continued)

(continued)

Year	Publication	Focus	Origin
5	Born, P. (2018). Regarding animals: A perspective on the importance of animals in early childhood environmental education. *International Journal of Early Childhood Environmental Education, 5*(2), 46–57	Content Animals in ECEE Post-humanism	USA
6	Jickling, B., Blenkinsop, S., Morse, M., & Jensen, A. (2018). Wild pedagogies: Six initial touchstones for early childhood environmental educators. *Australian Journal of Environmental Education, 34*(2), 159–171. https://doi.org/10.1017/aee.2018.19	Pedagogy Wild pedagogies and nature play	International
7	Jørgensen, N. J., Madsen, K. D., & Læssøe, J. (2018). Waste in education: The potential of materiality and practice. *Environmental Education Research, 24*(6), 807–817. https://doi.org/10.1080/13504622.2017.1357801	Content Waste in education Materiality	Denmark
8	Kahriman-Pamuk, D., & Olgan, R. (2018). Teacher practices and preschool physical environment for education for sustainable development: Eco vs. ordinary preschools. Mersin University *Journal of the Faculty of Education, 14*(2), 669–683	Pedagogy Teacher practices and preschool physical environments	Türkiye
9	Moore, D., Almeida, S., & Barnes, M. (2018). Education for sustainability policies: Ramifications for practice. *Australian Journal of Teacher Education, 43*(11), 105–121	Education for sustainability policies Curriculum policy	Australia
10	Nxumalo, F. (2018). Stories for living on a damaged planet: Environmental education in a preschool classroom. *Journal of Early Childhood Research, 16*(2), 148–159. https://doi.org/10.1177/1476718X17715499	Content Post-humanism	Canada
11	Rooney, T. (2018). Weather worlding: Learning with the elements in early childhood. *Environmental Education Research, 24*(1), 1–12	Content Post-humanism, post-developmental research Ethnography	Australia

(continued)

(continued)

Year	Publication	Focus	Origin
2019			
1	Ärlemalm-Hagsér, E., & Larsson, J. (2019). Between the global and the local–preschool teacher students' meaning-making on Education for Sustainability in Swedish early childhood education. *NorDiNa, 15*(4), 370–386. https://doi.org/10.5617/nordina.6212	Teacher education Student teacher meaning-making	Sweden
2	Bascopé, M., Perasso, P., & Reiss, K. (2019). Systematic review of education for sustainable development at an early stage: Cornerstones and pedagogical approaches for teacher professional development. *Sustainability, 11*(3), 719. https://doi.org/10.3390/su11030719	Teacher development Systematic review of education for sustainable development in early education	International
3	Borg, F., Winberg, T. M., & Vinterek, M. (2019). Preschool children's knowledge about the environmental impact of various modes of transport. *Early Child Development and Care, 189*(3), 376–391. https://doi.org/10.1080/03004430.2017.1324433	Children's learning and understandings Preschool children's knowledge	Sweden
4	Boldermo, S & Ødegaard, E. E. (2019). What about the migrant children? The state-of-the-art in research claiming social sustainability. *Sustainability, 11*(1), 459. https://doi.org/10.3390/su11020459	Content Systematic review migrant children Social sustainability	International
5	Boyd, D. (2019). Utilising place-based learning through local contexts to develop agents of change in early childhood education for sustainability. *Education 3–13, 47*(8), 983–997. https://doi.org/10.1080/03004279.2018.1551413	Pedagogy Place-based learning	UK
6	Buil, P., Roger-Loppacher, O., & Tintoré, M. (2019). Creating the habit of recycling in early childhood: A sustainable practice in Spain. *Sustainability, 11*(22), 6393. https://doi.org/10.3390/su11226393	Content areas Creating habits for recycling	Spain

(continued)

(continued)

Year	Publication	Focus	Origin
7	Ernst, J., & Burcak, F. (2019). Young children's contributions to sustainability: The influence of nature play on curiosity, executive function skills, creative thinking, and resilience. *Sustainability, 11*(15), 4212. https://doi.org/10.3390/su11154212	Children's contributions Nature play	International
8	Heggen, M. P., Sageidet, B. M., Goga, N. et al. (2019) Children as eco-citizens? *NorDiNa: Nordic studies in science education, 15*(4), pp. 387–402. https://doi.org/10.5617/nordina.6186	Children's understanding and learning Citizenship, children, and adults being and becoming eco-citizens	Norway
9	Hirst, N. (2019). Education for sustainability within early childhood studies: Collaboration and inquiry through projects with children. *Education 3–13, 47*(2), 233–246. https://doi.org/10.1080/03004279.2018.1430843	Children's understanding and learning Projects with children	UK
10	Inoue, M., Elliott, S., Mitsuhashi, M., & Kido, H. (2019). Nature-based early childhood activities as environmental education? A review of Japanese and Australian perspectives. *Japanese Journal of Environmental Education, 28*(4), 21–28	Pedagogy Nature play, forest school	International
11	Kahriman-Pamuk, D., Uzun, N. B., Yıldız, T. G., & Haktanır, G. (2019). Reliability of indicators measuring early childhood education for sustainability: A study in Turkey using generalizability theory. *International Journal of Early Childhood, 51*(2), 193–206	Measuring early childhood education for sustainability	Türkiye
12	Lee, R. E., Soltero, E. G., Ledoux, T. A., Sahnoune, I., Saavadra, F., Mama, S. K., & McNeill, L. H. (2019). Sustainability via active garden education: Translating policy to practice in early care and education. *The Journal of School Health, 89*(4), 257–266. https://doi.org/10.1111/josh.12734	Pedagogies Garden pedagogy	USA

(continued)

(continued)

Year	Publication	Focus	Origin
13	Mahat, H., Hashim, M., Saleh, Y., Nayan, N., & Norkhaidi, S. B. (2019). Environmental sustainability knowledge, attitude and practices among pre-school students. *IOP Conference Series. Earth and Environmental Science*, *286*, 12,003. https://doi.org/10.1088/1755-1315/286/1/012003	Teacher education Knowledge, attitude, and practices among preschool students	Malaysia
14	Pamuk, S., & Kahriman-Pamuk, D. (2019). Preservice teachers' intention to recycle and recycling behaviour: The role of recycling opportunities. *International Electronic Journal of Environmental Education*, *9*(1), 33–45	Content Recycling	Türkiye
15	Sahin, E., & Alici, Ş. (2019). An associational study on pre-service early childhood teachers' nature relatedness in education for sustainability. *International Research in Geographical and Environmental Education, 28*(4), 343–357. https://doi.org/10.1080/10382046.2019.1651040	Teacher's understandings Nature relatedness	Türkiye
16	Sundberg, B., Areljung, S., & Ottander, C. (2019). Opportunities for education for sustainability through multidimensional preschool science. *Nordina: Nordic Studies in Science Education, 15*(3), 358–369. https://doi.org/10.5617/nordina.6237	Content Multidimensional preschool science	Sweden
17	Türkoğlu, B. (2019). Opinions of preschool teachers and pre-service teachers on environmental education and environmental awareness for sustainable development in the preschool period. *Sustainability*, *11*(18), 4925. https://doi.org/10.3390/su11184925	Teacher understandings Preschool teachers and pre-service teachers	Türkiye
18	You, S., Kim, E., & Shin, K. (2019). Teachers' belief and efficacy toward inclusive education in early childhood settings in Korea. *Sustainability*, *11*(5), 1489. https://doi.org/10.3390/su11051489	Teacher understandings Teachers' beliefs and efficacy	South Korea

(continued)

(continued)

Year	Publication	Focus	Origin
2020			
1	Ardoin, N., & Bowers, A. (2020). Early childhood environmental education: A systematic review of the research literature. *Education Research Review*, *31*, 1–16. https://doi.org/10.1016/j.edurev.2020.100353	Methodological and theoretical perspectives Systematic review	USA
2	Boyd, D. (2020). The legacy café - the holistic benefit of reviving lost arts, crafts and traditional skills through an early childhood intergenerational sustainable skills project. *Education 3–13*, *48*(2), 204–216. https://doi.org/10.1080/03004279.2019.1603247	Pedagogy Intergenerational sustainable skills project	UK
3	Fonsén, E., & Soukainen, U. (2020). Sustainable pedagogical leadership in Finnish early childhood education (ECE): An evaluation by ECE professionals. *Early Childhood Education Journal*, *48*(2), 213–222. https://doi.org/10.1007/s10643-019-00,984-y	Pedagogical leadership	Finland
4	Jørgensen, N. J., & Martiny-Bruun, A. (2020). Painting trees in the wind: Socio-material ambiguity and sustainability politics in early childhood education with refugee children in Denmark. *Environmental Education Research*, *26*(9–10), 1406–1419. https://doi.org/10.1080/13504622.2019.1602755	Pedagogy Refugee children New materialist/'common world' approaches and ethnography	Denmark
5	Melis, C., Wold, P.-A., Bjørgen, K., & Moe, B. (2020). Norwegian kindergarten children's knowledge about the environmental component of sustainable development. *Sustainability*, *12*(19), 8037. https://doi.org/10.3390/su12198037	Children's learning and understandings Children's knowledge	Norway
6	Powell, S., & Somerville, M. (2020). Drumming in excess and chaos: Music, literacy and sustainability in early years learning. *Journal of Early Childhood Literacy*, *20*(4), 839–861. https://doi.org/10.1177/1468798418792603	Content Music, literacy	Australia

(continued)

(continued)

Year	Publication	Focus	Origin
7	Rahmayanti, H., Oktaviani, V., & Syani, Y. (2020). Development of sorting waste game android based for early childhood in environmental education. *Journal of Physics*. Conference Series, *1434*(1). https://doi.org/10.1088/1742-6596/1434/1/012029	Content Waste sorting game	Indonesia
8	Rodríguez-Marín, F., Puig Gutiérrez, M., López-Lozano, L., & Guerrero Fernández, A. (2020). Early childhood preservice teachers' view of socio-environmental problems and its relationship to the sustainable development goals. *Sustainability*, *12*(17), 7163. https://doi.org/10.3390/su12177163	Teacher understandings Technology and EE	Indonesia
9	Skarstein, T. H., & Skarstein, F. (2020). Curious children and knowledgeable adults - early childhood student-teachers' species identification skills and their views on the importance of species knowledge. *International Journal of Science Education*, *42*(2), 310–328. https://doi.org/10.1080/09500693.2019.1710782	Teacher education Student teacher understandings of species	Norway
10	Tzima, S., Styliaras, G., Bassounas, A., & Tzima, M. (2020). Harnessing the potential of storytelling and mobile technology in intangible cultural heritage: A case study in early childhood education in sustainability. *Sustainability*, *12*(22), 9416. https://doi.org/10.3390/su12229416	Content Storytelling and mobile technology	Greece
11	Weldemariam, K. (2020). "Becoming-with bees": Generating affect and response-abilities with the dying bees in early childhood education. *Discourse*, *41*(3), 391–406. https://doi.org/10.1080/01596306.2019.1607402	Content Post-anthropocentric approaches Assemblage pedagogies	Sweden

(continued)

(continued)

Year	Publication	Focus	Origin
12	Wolff, L.-A., Skarstein, T. H., & Skarstein, F. (2020). The mission of early childhood education in the Anthropocene. *Education Sciences, 10*(2), 27. https://doi.org/10.3390/educsci10020027	Pedagogy ECE in the Anthropocene	International
2021			
1	Ampartzaki, M., Kalogiannakis, M., & Papadakis, S. (2021). Deepening our knowledge about sustainability education in the early years: Lessons from a water project. *Education Sciences, 11*(6), 251–251. https://doi.org/10.3390/educsci11060251	Content Water project	Greece Cyprus Bulgaria Bosnia Herzegovina Croatia Poland Iceland
2	Bergan, V., Krempig, I. W., Utsi, T. A., & Bøe, K. W. (2021). I want to participate—Communities of practice in foraging and gardening projects as a contribution to social and cultural sustainability in early childhood education. *Sustainability, 13*(8), 4368. https://doi.org/10.3390/su13084368	Content Gardening projects	Norway
3	Borg, F., & Gericke, N. (2021). Local and global aspects: Teaching social sustainability in Swedish preschools. *Sustainability, 13*(7), 3838. https://doi.org/10.3390/su13073838	Teacher understandings and development Social sustainability	Sweden
4	Engdahl, I., Pramling Samuelsson, I., & Ärlemalm-Hagsér, E. (2021). Swedish teachers in the process of implementing education for sustainability in early childhood education. *New Ideas in Child and Educational Psychology, 1*(1), 3–23	Teacher's understanding and learning Implementing education for sustainability	Sweden
5	Ernst, J., McAllister, K., Siklander, P., & Storli, R. (2021). Contributions to sustainability through young children's nature play: A systematic review. *Sustainability, 13*(13), 7443. https://doi.org/10.3390/su13137443	Pedagogy Systematic review of young children's nature play	International
6	Furu, A.-C., Kaihovirta, H., & Ekholm, S. (2021). Promoting sustainability through multimodal storytelling. *Journal of Teacher Education for Sustainability, 23*(2), 18–29. https://doi.org/10.2478/jtes-2021-0014	Teacher education Multimodal storytelling	Finland

(continued)

(continued)

Year	Publication	Focus	Origin
7	Güler Yıldız, T., Öztürk, N., İlhan İyi, T., Aşkar, N., Banko Bal, Ç., Karabekmez, S., & Höl, Ş. (2021). Education for sustainability in early childhood education: A systematic review. *Environmental Education Research, 27*(6), 796–820. https://doi.org/10.1080/13504622.2021.1896680	Methodological and theoretical perspectives Systematic review Education for sustainability in ECE	International
8	Hedefalk, M., Caiman, C., Ottander, C., & Almqvist, J. (2021). Didactical dilemmas when planning teaching for sustainable development in preschool. *Environmental Education Research, 27*(1), 37–49. https://doi.org/10.1080/13504622.2020.1823321	Teacher students' didactic choices	Sweden
9	Huda, M., Mustafa, M. C., & Mohamed, A. K. (2021). Understanding of multicultural sustainability through mutual acceptance: Voices from intercultural teachers' previous early education. *Sustainability, 13*(10), 5377. https://doi.org/10.3390/su13105377	Teacher understandings Multicultural sustainability	Malaysia
10	Johansson, E., & Rosell, Y. (2021). Social sustainability through children's expressions of belonging in peer communities. *Sustainability, 13*(7), 3839. https://doi.org/10.3390/su13073839	Content Social sustainability and belonging Preschool children and peer community	Norway
11	Ødegaard, E. E. (2021). Reimagining "collaborative exploration"—A signature pedagogy for sustainability in early childhood education and care. *Sustainability, 13*(9), 5139. https://doi.org/10.3390/su13095139	Pedagogy Collaborative exploration Social sustainability	Norway
12	Sadownik, A. R., & Gabi, J. (2021). (Re)imagining entangled sustainability: A human and nonhuman theorisation of belonging to safeguard sustainability's holism. *Sustainability, 13*(9), 4714. https://doi.org/10.3390/su13094714	Pedagogy Human and nonhuman theorising	International
13	Simsar, A. (2021). Young children's ecological footprint awareness and environmental attitudes in Turkey. *Child Indicators Research, 14*(4), 1387–1413. https://doi.org/10.1007/s12187-021-09,810-7	Children's understanding Children's ecological footprint awareness	Türkiye

(continued)

(continued)

Year	Publication	Focus	Origin
14	Spiteri, J. (2021a). Why is it important to protect the environment? Reasons presented by young children. *Environmental Education Research*, *27*(2), 175–191. https://doi.org/10.1080/13504622.2020.1829560	Children's understanding Environmental education and children's reasons for protecting the environment	Malta
15	Spiteri, J. (2021b). Can you hear me? Young children's understanding of environmental issues. *International Studies in Sociology of Education*, *30*(1–2), 191–213. https://doi.org/10.1080/09620214.2020.1859401	Children's understandings Children's understandings of environmental issues	Malta
2022			
1	Borg, F., & Samuelsson, I. P. (2022). Preschool children's agency in education for sustainability: the case of Sweden. *European Early Childhood Education Research Journal*, *30*(1), 147–163. https://doi.org/10.1080/1350293X.2022.2026439	Preschool children's agency	Sweden
2	Caiman, C., Hedefalk, M., & Ottander, C. (2022). Pre-school teaching for creative processes in education for sustainable development—Invisible animal traces, purple hands, and an elk container. *Environmental Education Research*, *28*(3), 457–475. https://doi.org/10.1080/13504622.2021.2012130	Pedagogy Teaching creative processes	Sweden
3	Chapman, S. N., & O'Gorman, L. (2022). Transforming learning environments in early childhood contexts through the arts: Responding to the United Nations Sustainable Development Goals. *International Journal of Early Childhood*, *54*(1), 33–50. https://doi.org/10.1007/s13158-022-00,320-3	Content Transforming learning environments in early childhood Art education	Australia
4	Feriver, S. (2022). Pre-schoolers as systems thinkers: Testing the water. *Environmental Education Research*, *28*(3), 430–456. https://doi.org/10.1080/13504622.2021.2002820	Children's systems thinking Preschooler systems education and assessment	Türkiye

(continued)

(continued)

Year	Publication	Focus	Origin
5	Raivio, M., Skaremyr, E., & Kuusisto, A. (2022). Caring for worldviews in early childhood education: theoretical and analytical tool for socially sustainable communities of care. *Sustainability, 14*(7), 3815–https://doi.org/10.3390/su14073815	Methodological and theoretical perspectives Socially sustainable communities of care Feminist and post-colonial ethics of care	Sweden
6	Spiteri, J., Higgins, P., & Nicol, R. (2022). It's like a fruit on a tree: Young Maltese children's understanding of the environment. *Early Child Development and Care, 192*(7), 1133–1149. https://doi.org/10.1080/03004430.2020.1850444	Children's understanding of the environment	Malta
7	Wakefield, W., Weinberg, A. E., Pretti, E., Merritt, E. G., & Trott, C. (2022). "When I act consciously, I can see a brighter world around me": Preservice teacher readiness to support transformative sustainability learning. *Environmental Education Research, 28*(11), 1672–1690. https://doi.org/10.1080/13504622.2022.2085246	Pre-service teachers Education as transformative Sustainability learning Environmental education	USA
8	Weldemariam, K. (2022). Broadening ontological and epistemological possibilities within early childhood teacher education for sustainability. *Australian Journal of Teacher Education, 47*(5), 33–49. https://doi.org/10.14221/ajte.2022v47n5.3	Teacher education Post-humanism	Sweden
9	Weldemariam, K., Chan, A., Engdahl, I., & Pramling Samuelsson, I. (2022). Care and social sustainability in early childhood education: Transnational perspectives. *Sustainability, 14*(9), 4952. https://doi.org/10.3390/su14094952	Content Country comparisons Care and social sustainability Transnational perspectives	International
10	Willockx, D., & Dom, L. (2022). Childcare 2030: On using SDGs as a framework for implementing sustainability in two day-care centres in Flanders. *International Journal of Early Childhood, 54*(1), 75–91. https://doi.org/10.1007/s13158-022-00,316-z	Pedagogies SDGs as a framework for sustainability	Belgium

(continued)

(continued)

Year	Publication	Focus	Origin
11	Višnjic-Jevtic, A., Lepicnik Vodopivec, J., Pribišev Beleslin, T., & Šindic, A. (2022). Unmasking sustainability in early childhood education: Teachers' voices from Bosnia and Herzegovina, Croatia, and Slovenia. *International Journal of Early Childhood, 54*(1), 119–137. https://doi.org/10.1007/s13158-022-00,321-2	Teacher understandings Teachers' voices	Bosnia, Herzegovina, Croatia and Slovenia

References

Almeida, S., & O'Hara. (2020). Research in early childhood education for sustainability: Policies and perspectives from India. In J. Davis & S. Elliott (Eds.), *Research in early childhood education for sustainability: International perspectives and provocations* (pp. 82–93). Routledge.

Ärlemalm-Hagsér, E., Berg, B., & Sandberg, A. (2018). Economic sustainability in Swedish preschools. Preschool teachers and preschools as political-economic actors. *Utbildning & Demokrati, 27*(2), 15–36.

Ärlemalm-Hagsér, E., & Elliott, S. (2021). *Oxford Bibliographies: Outdoor play and learning in early childhood education.* Oxford University Press.

Ärlemalm-Hagsér, E., & Pramling Samuelsson, I. (2021). "Business as usual"? Or transformative and transactive teaching leading towards the Agenda 2030 goals in Swedish Early Childhood Education. *International Journal of Early Childhood Environmental Education, 9*(1), 94–111.

Australian Government Department of Education (AGDE). (2022). *Belonging, being and becoming: Early years learning framework for Australia,* V 2.0. Australian Government Department of Education for the Ministerial Council.

Borg, F. (2017). Kids, cash and sustainability: Economic knowledge and behaviors among preschool children. *Cogent Education, 4*(1).

Christensen, M. (2021). *Children's understandings of sustainability related topics and issues: A phenomenographic investigation seen through drawings and interviews with 6-8-year-old children.* Professional Doctorate thesis, QUT, Australia.

Christensen, M., O'Gorman, L., Walsh, K., & Davis, J. (2024). Children's understandings of sustainability topics and the 4C pedagogical Framework. In J. Davis., & S. Elliott (Eds.), *Young children and the environment: Early education for sustainability.* Cambridge University Press.

Davis, J. (2009). Revealing the research 'hole' of early childhood education for sustainability: A preliminary survey of the literature. *Environmental Education Research, 15*(2), 227–241. https://doi.org/10.1080/13504620802710607

Davis, J. (2014). Examining early childhood education through the lens of education for sustainability: Revisioning rights. In J. Davis & S. Elliott (Eds.), *Research in early childhood education for sustainability: International perspectives and provocations* (pp. 21–37). Routledge.

Davis, J., & Elliott, S. (2003). Early childhood environmental education: Making it mainstream. *Early Childhood Australia.*

Davis, J., & Elliott, S. (Eds.). (2014). *Research in early childhood education for sustainability: International perspectives and provocations.* Routledge.

Davis, J. (2022). *Oxford Bibliographies: Sustainability in early childhood education.* Oxford University Press.

Dean, S., & Elliott, S. (2022). Urgency, equity, and agency: An assemblage of global concerns and interests in education for sustainability. *International Journal of Early Childhood Environmental Education., 9*(2), 56–68.

Elliott, S., & Ärlemalm-Hagsér, E. (2019). Early childhood - sociocultural contexts: Sustainability. In M. A. Peters (Ed.), *Encyclopedia of teacher education.* Springer. https://doi.org/10.1007/978-981-13-1179-6_101-1

Elliott, S., & Davis, J. (2018). Challenging taken-for-granted ideas in early childhood education: A critique of Bronfenbrenner's ecological systems theory in the age of post-humanism. In K. Malone, E. Barratt Hacking, & A. Cutter-Mackenzie (Eds.), *Research handbook on childhood-nature: Assemblages of childhood and nature research (Springer international handbooks of education)* (pp. 1–36). Springer.

Elliott, S., & Young, T. (2016). Nature by default in early childhood education for sustainability. *Australian Journal of Environmental Education, 32*(1), 57–64. https://doi.org/10.1017/aee.2015.44

Emery, S., Davis, J., Sageidet, B., Hirst, N., Boyd, D., & Browder, J. (2017). Transnational dialogues for sustainability research in early childhood education: a model for building capacity for ESD in universities? In Filho, W. L., Rogers, J., Castro, P., do Paco, A., Skanavis, C., & Kuznetsova, O. (Eds.), *Handbook of theory and practice of sustainable development in higher education, Volume 2 [World Sustainability Series]* (pp. 143–156). Springer.

Engdahl, I. P., Samuelsson, I., & Ärlemalm-Hagsér, E. (2023). Systematic child talks in early childhood education—A method for sustainability. *Children, 10*, 661. https://doi.org/10.3390/children10040661

Green, C. J. (2015). Toward young children as active researchers: A critical review of the methodologies and methods in early childhood environmental education. *Journal of Environmental Education, 46*(4), 207–229. https://doi.org/10.1080/00958964.2015.1050345

Hedefalk, M., Almqvist, J., & Östman, L. (2015). Education for sustainable development in early childhood education: A review of the research literature. *Environmental Education Research, 21*(7), 975–990. https://doi.org/10.1080/13504622.2014.971716

Harwood, D., Whitty, P., Green, C., & Elliot, E. (2020). In S. Elliott, E. Ärlemalm-Hagsér & J. Davis (Eds.), *Researching early childhood education for sustainability: Challenging assumptions and orthodoxies* (pp. 25–37). Routledge.

Hughes, F., Elliott, S., Anderson. K., & Chancellor, B. (2022). *Early years learning in Australian natural environments.* Oxford University Press.

Hughes, F. (2021). *Problematising sustainability in nature play programs: Pedagogical transformations and ambitious alignments* [Doctor of Philosophy, University of New England].

Inoue, M. (2014). Perspectives on early childhood environmental education in Japan: Rethinking for a sustainable society. In J. Davis & S. Elliott (Eds.), *Research in early childhood education for sustainability: International perspectives and provocations* (pp. 79–96). Routledge.

Kim, S., & Dreamson, N. (2020). Culturally inclusive early childhood education for sustainability: A comparative document analysis between Australian and Korean curricula. *European Early Childhood Education Research Journal, 28*(5), 712–730.

Miller, M. (2014). Intercultural dialogues in early childhood education for sustainability: Embedding Indigenous perspectives. In J. Davis & S. Elliott (Eds.), *Research in early childhood education for sustainability: International perspectives and provocations* (pp. 63–78). Routledge.

Nelson, N., & Hodgins, D. (2020). In S. Elliott, E. Ärlemalm-Hagsér & J. Davis (Eds.), *Researching early childhood education for sustainability: Challenging assumptions and orthodoxies* (pp.150–165). Routledge.

Paujik, Y. M., Miller, M., Gibson, M., & Walsh, K. (2021). 'Doing' socio-political sustainability in early childhood: Teacher-as-researcher reflective practices. *Global Studies of Childhood, 11*(3), 265–280. https://doi.org/10.1177/2043610620941133

Ritchie, J. (2014). Learning from the wisdom of elders. In J. Davis & S. Elliott (Eds.), *Research in early childhood education for sustainability: International perspectives and provocations* (pp. 49–36). Routledge.
Ritchie, J. (2021). *Oxford Bibliographies: Sustainability education in early childhood education.* Oxford University Press.
Somerville, M., & Williams, C. (2015). Sustainability education in early childhood: An updated review of research in the field. *Contemporary Issues in Early Childhood, 16*(2), 102–117. https://doi.org/10.1177/1463949115585658
United Nations (UN). (1989). *United Nations Convention on the Rights of the Child* (UNCRC). www.ohchr.org/en/professionalinterest/pages/crc.aspx
Weldemariam, K., & Wals, A. (2020). From autonomous child to a child entangled within an agentic world: Implications for early childhood education for sustainability. In S. Elliott, E. Ärlemalm-Hagsér, & J. Davis (Eds.), *Researching early childhood education for sustainability: Challenging assumptions and orthodoxies* (pp. 13–24). Routledge.

Chapter 5
Enablers and Turning Points in ECEfS

Abstract In this chapter, we explore—beyond our individual and shared ECEfS histories—with invited international colleagues. As a contextually diverse group, these respondents offer unique insights drawn from their many years of participation in ECEfS. We thank them for generously responding to our questions to inform the drafting of this chapter. We analysed and interpreted their responses and then shared the chapter pre-publication for editorial comment. Thus, in this chapter, we offer our interpretation and synthesis of their insightful contributions, supported by text box provocations for readers about their own participation in ECEfS.

5.1 Introduction

Throughout this text, we have been explicit about our individual and shared histories in ECEfS. In this chapter, we recognise a small sample of the many international colleagues who have been interwoven with these histories and the broader evolution of ECEfS. We purposefully selected a group of ten academics and practitioners with long-term participation in ECEfS and invited them to critically reflect on and respond in writing via email to a set of questions about their own professional histories and perspectives (see Appendix 1). The intention was to draw on their contextually diverse and sustained participation in ECEfS research and practice. As co-authors informed by social constructivist theorising who see learning as a collaborative endeavour, we recognise that each respondent, like ourselves, has brought their unique perspective to this collective history. Therefore, this is not a typical book chapter, but one enriched by many voices. We collated, then thematically analysed their responses and cross-checked our co-author interpretations of emerging themes. We sought commonalities as well as unique insights informed by each respondents' socio-cultural context. At times, we engaged in minor editing for the purposes of reader clarity. All respondents

We dedicate this chapter to Glynne Mackey of Aotearoa New Zealand who sadly passed away in June 2023. Glynne was one of the earliest members of ECEfS. Her clarity of thinking, speaking and writing about topics related to children, environment and sustainability is already deeply missed by her many friends and colleagues in this movement.

J. M. Davis et al., *Early Childhood Education for Sustainability*,
SpringerBriefs in Education, https://doi.org/10.1007/978-981-97-8405-9_5

were invited to review the chapter pre-publication to edit any misinterpretations. Importantly, we acknowledge each person's generosity and trust in interweaving their stories with ours in writing this text. This chapter offers our interpretation and synthesis of their insightful contributions.

We have organised respondent reflections in the following way: initial triggers for becoming engaged in ECEfS; then key influences, personal contributions, and ongoing involvement in ECEfS. We continue with their perceptions about the wider field, discussing their ideas about enablers and constraints, pivotal turning points and future trajectories. Respondent commonalties and concerns, interwoven with our shared stories (see Chap. 2), build a bigger and stronger picture of the history of and possible futures for ECEfS.

5.2 Initial Triggers for ECEfS Participation

A diversity of initial triggers was reported for initial participation in ECEfS. Nevertheless, we were able to cluster these into three main categories: formative personal experiences in and about nature and environment; local, national and/or global professional connections; and broader societal triggers such as interest in political or cultural systems and values.

Regarding the first point, several respondents, including Mackey from Aotearoa/New Zealand, reflected on her formative childhood nature experiences:

> My commitment and passion for EfS has grown out of my childhood experiences ... many native bush walks, identifying plants with my botanist father, and camping. All this transferred into my work within teacher education programmes and realising the impact teachers could have on very young learners in the early years (GM).

Similarly, across the Pacific in the USA, McCrea cited growing up in rural California on a ranch enjoying homegrown produce as influential, alongside her early professional ECE roles, before moving to Australia in the mid-1970s. Also in the USA, Wilson recognised her '*personal connectedness with nature strongly influenced ... involvement in ECEE and ECEfS*'. She identified that '*most young children were more joyful, curious, and imaginative while outdoors in a natural environment*' and sought to intervene and address a perceived '*prejudice against nature*'. For others, initial studies and professional work in the natural sciences extended into environmental issues beyond simply '*nature-friendly topics*' *(OJ)* often typical in ECE. For example, Inoue from Japan stated:

> I realized that the nature I had learned about and experienced was completely different from the nature perceived by ECE professionals and pedagogues... A pedagogy that sees nature merely as a tool for children's development cannot provide education that leads to the protection of nature or sustainability. I began to think about the reason and how to change ECE, and then I started to engage in research on environmental education in early childhood (MI).

On the second point, professional connections at local, national, and/or global levels were important to several respondents. In 1992, Hocking, an Australian EC practitioner, joined a fledgling local network EEEC Vic Inc (see Chap. 2) and '*quickly realized we shared common values*'. The network '*provided a substantive base from which to implement the learnings I was undertaking around EE*' (MH). Moving from local to global, Ji from South Korea and Li from China described professional connections facilitated by international visits and exchanges as highly relevant. Li reflected on extended dialogues through her experiences of deep ecology in Norway and her own country's Daoism beliefs.

Growing up in Swedish society with a history of sustainability, Pramling Samuelsson recalled, '*even my grandmother washed the plastic bags, and people exchanged services with each other*'. This likely set the stage for her later leading global connections and involvement (also see Chap. 2).

> In 2003, Gothenburg University together with the Swedish Government arranged a conference entitled a Consultation about Education for Sustainable Development – this was a gigantic challenge. I had the opportunity to invite 10 researchers from various regions in the world who were interested in ECE. I also had an opportunity to appeal to the academic world at the closing conference ceremony about the place of the youngest children in ESD. This conference led to four international workshops, with one about ECE that I jointly arranged with UNESCO. This workshop resulted in the report, Pramling Samuelsson and Kaga (2008). In 2008, I became OMEP World President and resolved that peace was one aspect of sustainability and this is why OMEP became the first organization to focus on ESD for the youngest children. At the same, I was invited by the Swedish Government to become the UNESCO Chair in ECE and SD. (IPS)

Thirdly, personal interest in political or cultural systems and values were initial triggers for both Engdahl (Sweden) and Ritchie (Aotearoa/New Zealand). As an early childhood teacher, Engdahl was responding to '*political interest in environmental issues and the green transformative wave that swept through western societies in the 1970's*', while Ritchie drew on '*cultural sustainability in line with the Aotearoa (New Zealand) context of commitment to Te Tiriti o Waitangi*' (Treaty of Waitangi). Ritchie also described a critical trigger point during the drafting of a research funding application, when a participating teacher urged the prioritising of climate change. As Ritchie stated, '*this challenge resulted in a shift in our thinking, from manaakitanga, a focus on caring for other people, to include kaitiakitanga, the obligation to actively care for the environment*' (Ritchie et al., 2010). Thus, central to their proposed study was the '*intersectionality of social, cultural and environmental justice in early childhood care and education, and how te ao Māori (a Māori worldview) provides a platform for pedagogies in ECEfS and ESD in Aotearoa*'.

Text Box 5.1

We invite readers to reflect on their own triggers for involvement in ECEfS. Are there formative life stories, local, national or global professional connections, or perhaps systems and values core to your worldviews? Such triggers will likely inform how we move forward individually and collectively over coming decades to further provoke and shift sustainability thinking and action across the ECE field.

5.3 Key Influences Over Time: Authors, Colleagues and Organisations

Our respondents cited many influences, from early thinkers and authors through to contemporary colleagues in shaping their professional lives and ongoing interest in ECEfS as well as selected organisations and networks that aligned with their personal interests.

Significant early authors include Rachel Carson, Teilhard de Chardin, and Forrest Carter. Pioneering scientist and environmentalist Carson (1907–1964) is still often cited for her 1950s ground-breaking work, culminating in *Silent Spring* (1962). She pre-empted global predicaments warning that 'we are dealing with dangerous things and it may be too late to wait for positive evidence of danger' (p. 16). Less defiant, but in many ways more significant for ECE, was Carson's (1956, republished 1998) essay *A Sense of Wonder*, that documented her nature ramblings with her young nephew. This essay alerted educators to the foundational importance of children's connections with nature at a time when many children were quite free to roam in nature; many decades later, the situation is far more perilous.

de Chardin (1881–1955) was a French philosopher, theologian, and palaeontologist. He attempted to reconcile his Catholic beliefs with Darwinian evolution understandings and broader ideas about humans and the universe published in *The Divine Milieu* (1957, republished 2004). In referring to these influential authors, Wilson (USA) reflected:

> The writings of Rachel Carson and Teilhard de Chardin were strong influencers. Carson's Silent Spring raised my awareness about the extent to which humans were harming the environment; and, The Sense of Wonder reminded me of the importance of the natural world to our holistic well-being. Carson's comment about a "true instinct for what is beautiful and awe-inspiring" being "dimmed or even lost before we reach adulthood" motivated me to do something about keeping the sense of wonder alive in myself and in children. De Chardin's, The Divine Milieu, shifted my worldview. The shift was from viewing the world as a resource to understanding that the entire universe is holy. I wanted to honor this holiness.

Another respondent, South Korean, Ji, also recalled the influence of *The Sense of Wonder* (1956, republished 1998) and, in addition, cited Forrest Carter (1930–1987) who wrote *The Education of Little Tree* (1976), a fictional memoir about his Indigenous upbringing by Cherokee grandparents.

Beyond specific authors, there were many references to influential colleagues whose paths had intersected with our respondents. The following quotes from Hocking (Australia), Li (China), and Ritchie (Aotearoa/New Zealand) encapsulate how such intersections facilitated shared wisdom and expertise.

> Initially, I was influenced by the professional expertise of the Sue's - Sue Elliott, Sue Emmett and Sue Crook. Publications such as 'Snails Live in Houses Too' and 'Just Imagine' were the catalyst for significant change in my practice. The dozens of EC educators and outdoor play spaces I connected with through EEEC meetings at various locations around Melbourne, Geelong and beyond (MH).
>
> Prof. Julie Davis who introduced the concept of ECEfS to me in early 2010s, showed me the way forward. Prof. Åsta Birkeland who invited me to Norway, encouraged me to pave the way for ECEfS in China and connected our common ideas in different cultural contexts. Prof. Ingrid Pramling Samuelsson as a leading guest editor helped me to consider leadership as the ability to implement change (ML).
>
> So many influences, too many to list. With regard to understanding how colonization has contributed to environmental as well as human exploitation, the insights of the late Moana Jackson are an enduring source of wisdom. Working closely with Māori colleagues such as Cheryl Rau, Sandy Morrison and Mere Skerrett has provided deep insight. The pioneering ECEfS work of Julie Davis and Sue Elliott in Australia has been valuable (JR).

For Swedish respondents, there were organisations and networks that offered forums of broader mutual influence. Engdahl cites, *'I was active in our union for preschool teachers and in establishing a new Green party in Sweden'*, later participating in *'the first international workshop on ECE for ESD that took place in Gothenburg, co-hosted by the University of Gothenburg, Chalmers [University], and UNESCO Sweden'*. Both Engdahl and Pramling Samuelsson strongly referenced the role of the World Organisation for Early Childhood (OMEP) as a professional network with significant international reach and influence across 70 countries (see Chap. 2). Further, Pramling Samuelsson identified that in her role as UNESCO Chair for Early Childhood Education and Sustainable Development from 2008, she was able to build on the networks and projects instigated by OMEP thus creating 'a critical mass of influence'. Both OMEP and UNESCO are significant organisations internationally in promoting ECEfS (see Chap. 2).

Mackey offered an insight about the common bond shared by those seeking to influence or to be influenced; she saw this as 'breaking new ground' together. As a New Zealand university teacher educator and participant in the UNESCO Initial Teacher Education Network for Sustainability to reorient teacher education (Mackey & Ritchie, 2012), she stated:

> The strongest influences have come from fellow academics and graduating students who are also trying to break new ground within their institutions and bringing sustainability and social justice to the fore in their programmes ... [also] international academics who understand the valuable opportunity that sits within early childhood programmes to promote EfS (GM).

Text Box 5.2

Readers may reflect here on their own key influences over time. Was it a notable author, an unexpected, shared intersection with an individual, or perhaps, a bigger organisational stage? The key, we suggest, is being responsive to these influences, whether critically examining them, embracing them whole-heartedly or perhaps somewhere in between.

5.4 Respondents' Personal Contributions to ECEfS

The extensive personal contributions offered by the respondents are best collated and shared across three key areas: advocacy, organisational leadership, and EC teacher education.

Firstly, around advocacy, various respondents cited multiple authorships, research studies, projects, and their own explicit interventions to facilitate policy change for ECEfS. For example, Wilson advocated curriculum policy shifts when understandings of ECEE—as opposed to nature education—were limited in the USA. Mackey led a submission for policy change in New Zealand teacher education standards, and Hocking in Australia sought revised national health and safety policy to include sustainability priorities.

> In the early 1990's, there were only a few nature-based preschools in the United States, and funding for environmental education at the early childhood level was being questioned as to its legitimacy. Many people had little understanding about ECEE, yet I convinced a panel of reviewers to fund the first Ohio environmental education grant focusing on preschool-age children. (RW)

> In 2017, the Teaching Council of New Zealand/Aotearoa published the code of ethics and standards for all registered teachers to be assessed against, in order to maintain their registration status. As part of the consultation process, I led a submission from the University of Canterbury for sustainability and social justice to be prominent in the document. All teachers in Aotearoa/New Zealand are now required to demonstrate through their teaching, their responsibility to society in maintaining standards on sustainability, social justice and human rights. (GM)

> The behind the scenes input into Staying Healthy in Child Care 5th edition, when it first acknowledged sustainability perspectives (albeit sparingly). (MH)

These examples point to the much-needed, multidirectional and systemic policy shifts necessary to embed broader change towards sustainability in the ECE field.

Other respondents advocated for the uptake of ECEfS through a range of projects and initiatives at local, national, and international levels. Significantly, some of these have become long standing and much-reported internationally (Davis & Elliott, 2014; Engdahl, 2015; Siraj-Blatchford et al., 2010a, 2010b) offering inspiration for educators, policymakers and researchers alike. Ji and Engdahl outline their contributions below:

> My main contribution was the development of the 'Green Growth Education Program for Young Children' in 2012 with financial support from the Korean Ministry of Education and informed by ESD. Then, I educated the early childhood teachers including directors

nationwide. As a result, the program was applied to all kindergartens in Korea from 2013. (OJ)

Since 2010, the OMEP ESD projects have enrolled 142,491 young children, 15,574 teachers, 13,000 families and 1000 student teachers in EfS, expanding into eight foci:

- Children's understandings about sustainability and promoting change;
- Preschool practices around the 7 Rs;
- Intergenerational dialogues for sustainability;
- ESD for equity;
- Pre/Inservice teacher training;
- Environmental Rating Scale for Sustainable Development in Early Childhood, informed by children's knowledge, agency, and rights;
- As a key partner in the UNESCO Global Action Programme (GAP), OMEP opened a resource bank (http://ececresourcebank.org/index.php); and
- OMEP organizes an annual ESD in ECE award competition among 70 countries. (IE)

Innovative local/national projects are ongoing, for example, Li reported on the *'first demonstration ECEfS program created in a Shenzhen kindergarten in 2020, funded by Guangdong Province, China, with academic supports from our research team'*.

Secondly, several respondents were in organisational leadership roles that offered public platforms and a voice for far-reaching contributions. As President of OMEP Sweden, Engdahl had opportunities to be *'a project leader and coordinator, enabling teachers and children all over the world to engage in ECEfS, through articles, reports, and as a keynote speaker'*. Alongside Pramling Samuelsson, OMEP World President from 2008, a productive leadership team formed:

In 2010, we invited colleagues in ECE to an OMEP World Conference with the main theme Education for Sustainable Development, which was the first international conference on ESD for ECE. To gather not only adult perspectives, but also children's perspectives, OMEP Sweden initiated an World Project on ESD which has been ongoing ever since. I was the Congress Chair, and I have been the World Project leader for OMEP for many years, always collaborating closely with Ingrid Pramling Samuelsson, and initially with Professor Milada Rabušicová, from the Czech Republic (IE).

Further, from Pramling Samuelsson's perspective:

I think I have been a key person for putting ESD on the agenda for young children, not the least through the work we have done in OMEP ... participating in a number of UNESCO conferences and ... finally in the lobbying for Goal 4.2 in Agenda 2030. It is a continuous effort to cooperate with UNESCO.

As well as the more obvious global stage, there are those who have led ECEfS international delegations across organisations including Li from China to Norway, and Ji from South Korea to Australia. Exemplifying such exchanges, Li reported that her Chinese-Norwegian research team published a Special Journal Issue, *Building a Sustainable Future for Our Young Children in China and Nordic Countries* (Pramling Samuelsson et al., 2019). Inoue stated that, in her experience, fundamental to these exchanges was offering non-Western meanings or *'Japanese perspectives to Western-centred academics*'. Such exchanges provide fertile ground for cross-cultural, contextualised exchanges around key terms and meanings for both organisations and individuals (see Chap. 2). As Western academics and co-participants in several of these international exchanges, we have become increasingly aware of the multiple, and often contested, meanings of terms such as sustainability and child agency. Increasingly, we critically reflect on dominant Western approaches to ECEfS and seek to be open and inclusive. Nevertheless, in all these varied leadership roles, as Pramling Samuelsson states, our goal must be '*to be the voice of the youngest children in all these conferences and meetings*', advocating for their sustainable futures.

Lastly, most of the respondents, like ourselves, were involved in early childhood teacher education at pre-service and in-service levels; one suspects (and hopes!) that generations of educators have been influenced by these committed individuals. As Mackey stated: '*I place a high importance on early childhood teacher education programmes and how well we prepare beginning teachers to take sustainability to the heart of their teaching … to spread the message wider*'. Similarly, McCrea (Australia) was committed to *'educators' professional learning about EE/EfS'* offering many illustrative examples of educator-related projects and writings. These included topics such as: food cycles in the community in 1979; consuming lifestyles in 2005; a yearlong local ECE Eco-mentors program in 2012; and a leaders' ethic of eco-caring project in 2015. Building on her long-standing commitment and experience, she contributed a legacy, *Leading and Managing Early Childhood Settings: Inspiring people, places and practices* (McCrea, 2015) underpinned by an ethos of sustainability.

A final sentiment shared by Ritchie makes an explicit link between teacher education and research. She states, '*I would hope that our research has been accessible to teachers and has inspired them to understand ways in which they can apply a commitment to social, cultural and ecological justice in their work' (JR).* In line with these sentiments, Evans et al. (2022) have explored the critical importance of teacher education in ECEfS in a Special Issue of the *Australian Journal of Teacher Education.* Key findings on this topic alert to the necessity of continuing to address the ECEfS gap in teacher education highlighted by Davis and Davis (2020).

Text Box 5.3

We invite readers to reflect on what might be their contributions and legacies in years to come. This text is premised on the urgency of action by all, whether you are educators, researchers, governing organisational members, or policymakers. We ask, what is your priority now for leveraging change in the ECE field and what advocacy, leadership and teacher education knowledge and skills do you bring to this ongoing task to ensure your profound and lasting contributions to ECEfS.

5.5 Drivers for Ongoing Participation in ECEfS

We invited our respondents, many retired or close to retirement, to share what maintained their ongoing involvement in ECEfS. Responses varied, from long-held beliefs and values about the Earth and human impacts, to continued involvement in more pragmatic research projects and initiatives. However, an enduring commonality for all was the strength of relational commitments with colleagues, peak organisations and networks.

Wilson in the USA and McCrea in Australia, for example, shared motivations grounded in their beliefs and values, though differently expressed:

> My continued involvement with ECEE and ECEfS is motivated by my strong belief that caring for the environment is our best path to living the life we are meant to live. I believe our physical, emotional, and spiritual connections with the natural world call us to appreciate, be engaged with, and care for that which maintains life on Earth. (RW)

> My concerned motivation is the on-going state of the Earth alongside human actions: plastic, plastic, plastic and over-packaging everywhere; throw-away consumerism; damaged landscapes; ultra-processing of foods; sociocultural injustices; etc… all involving families; local communities; multi-levels of government and education systems; organisations. (NM)

More succinctly, Li (China) described a *'personal passion for a better world'* and Inoue (Japan) cited *'a sense of urgency/crisis about the future'*. There is no doubt from our reading of their responses that all shared a commitment to global sustainability and children's futures, if not always explicitly stated.

From the more pragmatic stance, several respondents described ongoing projects that maintained their engagement, whether leading research projects, writing, presenting, or guiding new program initiatives. Now, as elders in the field, their wisdom and expertise is highly valued.

> I am still engaged in OMEP's ESD awards and involved in a development and research project in Sweden called Sustainable Preschool that will last until the end of 2024. I am also participating in a research project on eco-certified preschools' meaning for children's learning about sustainability. (IPS)

> In September 2022, I opened the new 'EfS Early Childhood and Environment Research Institute'. Through this institute from March 2023, I plan to implement the ECEfS program that I originally developed with 10 ECE institutions that have different situations and conditions. I am going to provide guidance. (OJ)

> ESD and children's rights were important content in my position at Stockholm University, however, as a senior (semi-retired) professor I have now been able to initiate a few ECEfS research projects, for example, Swedish sustainable preschools projects and a European Erasmus+ project - Sustainability From the Start - an online free education program for ECE teachers. This is highly motivating, meeting and learning from so many engaged and involved teachers, leaders, and researchers, and being able to contribute with my competence in this field. (IE)

Core to many respondents' ongoing commitments were the strong relationships with colleagues both within and through peak organisations at international and local levels, such as OMEP, UNESCO, the TND, and EEEC (Victoria, Australia) (see Chap. 2). Mackey focused on the opportunities these relationships offered, *'research teams such as the Transnational Dialogues (TND) network and World OMEP have given me opportunities to be active in collaborative writing for research journals and book chapters'*. She also stated the *'commitment and passion of my fellow researchers is inspiring and motivating'*. As co-authors, we sense a wise and critical mass of people here, who thrive on the relationships and networking, informed by their shared beliefs and values.

Text Box 5.4
What keeps you involved and motivated in ECEfS, particularly if there are multiple challenges as outlined in earlier chapters and below? We suggest exploring local, national, and worldwide networks to find like-minded researchers and practitioners, and actively seek collegial engagement. Hopefully, this text will inspire you to not only maintain your participation, but to continue to build momentum for an even larger wise and critical mass of advocates, leaders, and change agents.

5.6 Reflecting on Key Enablers in the Evolution of This Field

The overriding theme to emerge in relation to enablers in ECEfS centred around networking among the heterogeneous stakeholders—from practitioners to academic researchers and policymakers. This is perhaps not surprising given the emphasis on the power of relationships for maintaining involvement stated above. Networking was described informally as *'talking with each other as colleagues ... [building] collaboration, a spirit of cooperation' (RW),* but also cited as a form of professional learning *'engaging in interpersonal recurrent 'Communities of Practice' with a mentor or two' (NM).* Engdahl aptly summed up the value of networking and networks thus:

> The best enabler to keep up the competence and involvement is networking, sharing research and ideas, inspiring and motivating each other. OMEP in Sweden, Europe and the World are important networks, and the TND research network has since 2010 become an important enabler for international collaboration. (IE)

In addition to the major international networks mentioned, networking was facilitated through conferences, invited presentations, research projects, or meetings at various levels. Mackey highlighted that *'national and international conferences that have a focus on education for sustainability and/or early childhood'* were key to her networking opportunities. A leader in South Korea, Ji created annual early childhood teacher conferences as an outcome from her case study research.

> I have been for many years an advisor and facilitator of an ECEfS learning community that consisted of around 16 members. Through this, we have made good project case studies

> related to ECEfS implementation like the 'Musim stream project'. As a next step, with these case studies, I held a big conference once a year to share the effective implementation of ECEfS with many teachers.... I think this method was a useful way to spread ECEfS far and wide. (OJ)

As a UNESCO Chair, Pramling Samuelsson had unique opportunities to capitalise on the value of conferences as ECEfS enablers on the world stage. She shared that *'the ESD work within UNESCO included the UN World Conferences in Moscow 2010 and in Uzbekistan 2022 and the Declarations that were accepted. (IPS).* Pramling Samuelsson also described an early *'lobby seminar in Brussels in 2010 together with the Swedish Center for Environment and Sustainability'* linked to a seminar report by Siraj-Blatchford et al., (2010a, 2010b). These illustrative points signal the potential of global declarations or government-led policy shifts and publications as strong motivators for advocacy and action.

In addition, both Ji and Ritchie have witnessed recent major policy changes in their respective countries and highlighted these as enablers for wider national uptake.

> The Korean government has been making visible efforts for carbon neutral education and climate change education since 2021. For example, in 2021 carbon neutral schools were selected by the Ministry of Education and received financial support, then in 2022, two public kindergartens were included. The Ministry of Education has now developed a Climate Change Education Program for young children and distributed this nationwide to kindergartens. (OJ)

> The openness and willingness of teachers in ECE services to enhance their teaching in ways that are consistent with obligations under Te Tiriti o Waitangi. … The Teaching Council of Aotearoa [since 2017] has required all teachers to adhere to a code of professional responsibility that includes:
>
> Code 4. Commitment to Society - I will respect my trusted role in society and the influence I have in shaping the future by:
>
> - Promoting and protecting the principles of human rights, sustainability and social justice
> - Demonstrating a commitment to Te Tiriti o Waitangi based Aotearoa New Zealand
> - Fostering learners to be active participants in community life and engaged in issues important to the wellbeing of society.
>
> (https://teachingcouncil.nz/professional-practice/our-code-our-standards/) (JR)

Some respondents linked their enablers to broader thinking about ECE, for example, *'ECE for the public good and sustainable development as a key value with qualified and quality teachers and informed parents' (ML).* As a counterpoint, Li's sentiments suggested that ongoing ECEfS uptake may be hindered without such a foundational enabler. In addition, Li and others offered enablers including the UNCRC (UNICEF, 1989) and a *'new view of young children' (MI)* as capable societal contributors. McCrea proposed, we can be enabling change with educators and children through *'nurturing repertoires of intentional rights with reciprocal EE/EfS responsibilities'.*

Lastly, McCrea and Pramling Samuelsson mentioned research and publications that offer evidence-based guidance and frameworks as important enablers for continued ECEfS uptake. McCrea also suggested that employing catch phrases may help to engage and guide practitioners, for example *'more-than-human'; 'think globally, act locally'*; and *'using one's head, heart, hands'*, given that some concepts in the field may initially be challenging. As previously documented (see Chaps. 2 and 4), research and publications have significantly increased and been shown to be influential enablers of transformative practice and policy.

There are also many enabling voices to be heard in ECEfS. Wilson offers the *'voices of writers, philosophers, poets, scientists, children, and even the Earth, itself'*, while Hocking cites the voices of *'committed individuals and organisations who have flown the flag for ECEfS in the form of interest groups, publications, research, resources and professional development'*. In sharing their ideas about enablers, our respondents demonstrated their own generosity of spirit as networkers and collaborators across multiple platforms and international borders. As co-authors, we endorse these social factors as essential to the evolution of ECEfS.

Text Box 5.5
A leading Australian practitioner, Hocking stated that she saw, as strong enablers, *'Educators, teachers, staff and boards of management of ECE centres that adopted a whole centre approach to EfS and then opened themselves to the wider ECE community to support, inspire and educate'*.

We invite you to emulate, in some way, the notions expressed here. How can you enable broad and systemic change within your immediate sphere of influence, while also being generous and open to the possibilities of supporting, inspiring and educating others? Just imagine the enabling potential of so many voices standing up for children's futures!

5.7 Reflecting on Key Constraints in the Evolution of This Field

Compelling themes emerged when we analysed the reported key constraints proffered by our respondents. Themes included: conceptual ambiguities around sustainability and nature; misguided views about ECE as well as children's rights and capabilities; neoliberal political agendas; and more pragmatically, persistent gaps in resourcing, training and curricula for ECEfS.

Respondents Ji, Inoue and Wilson, all shared their concerns about conceptual ambiguities around sustainability and nature. They perceived such ambiguities to be significant constraints, particularly when attempting to interpret curriculum policy and relevant learning for children, engaging in teacher education, or sharing views with a range of researchers. In forums bringing together a diversity of participants, it is apparent that one first needs to clarify key concepts—to ensure that everyone is somewhat 'on the same page', especially when aiming to be inclusive of non-Western understandings. We share these concerns as the varied international terms EE/ESD/EfS have also offered us challenges in drafting this text. To these conceptual ambiguities, we also add different perceptions of nature and its historical links to ECE theory and practice (see Chap. 2).

> Constraints include misunderstandings about ECEE and ECEfS. Misunderstandings relate to what children can and should be learning, as well as the way "nature" is viewed and defined. (RW)

> Ambiguity of differences between concepts: Terms such as 'ESD', 'EfS', and 'EE oriented to SD' are used simultaneously in the ECE field in Korea. (OJ)

> Constraints include various views of nature, especially differences between the views of natural scientists and those of social scientists. (MI)

Beyond such conceptual ambiguities, there is often a sense of marginalisation and pervasive misunderstandings about ECE and EfS. Respondents shared this constraint drawing on long-standing experience. In particular, Pramling Samuelsson stated *'one always has to be the voice of children ... they are often forgotten'*, while Engdahl made her concerns patently clear,

> Early childhood is never a priority, most often the concepts or content of early childhood are not even mentioned, not in the Agendas of the [UNESCO] meetings, nor in the publications. Early childhood is underfinanced, under researched and with low representation. (IE)

The compelling message is that the field needs to constantly advocate for young children and their healthy sustainable futures as a priority, as is increasingly supported by the research literature (see Chaps. 2 and 4).

Another key constraint is how young children can best be pedagogically engaged in co-creating sustainable futures when there are pervasive misunderstandings amongst educators and the broader community about young children their rights, roles, and capacities. In China, Li identified the commonly held '*image of children as vulnerable*' to be constraining when contemporary theorising and EfS approaches promote children as active and capable participants, an orientation not widely understood in China. Similarly, McCrea described *'limited sociocultural and ecological images of young children'* as a constraint. Such differently held or limited images about children can translate into misguided initiatives. From her experiences in South Korea, Ji offered an example of government instigated activities *'to attempt to teach difficult concepts related to ESD to young children directly'*:

> The Ministry of Environment and the Ministry of Education know the need for 'ESD for young children', so they have been doing various efforts nowadays. However, some activities might be not appropriate for the developmental level of young children and may lead to some misconceptions related to global warming. For example, as an activity for making children aware of global warming: Have young children draw a globe with special crayons (the original colour changes to a different colour when the temperature rises) and cut and paste it onto a paper cup. Then pour hot water into the paper cup and watch the colour of the picture of the globe pasted on the paper cup change. (OJ)

In this perplexing context of ambiguities and misunderstandings, there are also political systems and agendas that are constraining. Li described her concerns about a *'policy priority for a human capital orientation'*, while Ritchie pointed to the neoliberalisation of governments and the agendas this generates with negative impacts for ECE and EfS.

> In Aotearoa, the neoliberalisation of our education ministry has meant that there has been a lack of genuine consultation with ECE in the development of documents. There is also an ongoing serious lack of provision of funding for professional growth opportunities for teachers. In particular, it has failed to prioritise EfS or climate change education. ECE is largely privatised; and thus, has huge issues pertaining to quality, and sadly, many community-based, not-for-profit services have been unable to sustain their viability in the face of the power of the corporate chains. Without strong community-based services, it is much harder for ECEfS to be fostered, since this requires engagement in local community with local Māori input, and connections with protecting local biodiversity… not necessarily the priorities of a profit-oriented sector. (JR)

We now turn to the more pragmatic constraints reported by our respondents, specifically persistent gaps across resourcing, pre-service and in-service education, and curricula for ECEfS. The significant lack of ECE resourcing that Engdahl shared has been reiterated for Australia by Hocking. As an early convenor of the state-based network, EEEC Vic. Inc. (see Chap. 2), she highlighted funding, research, and material resourcing needs as constraining.

> Lack of government funding at a state and national level, lack of research, particularly in the early days – we knew in our hearts the value of green spaces and embedding sustainable practices in EC programs, but there was very little research and a lack of resources to support practice. Resources sold by EEEC in the 1990's went a long way towards filling this gap. Over time, many of the major commercial play material suppliers to [ECE] jumped on board the natural material bandwagon that continues to be [merely] a marketing tool. (MH)

Several respondents noted the limited and often ad hoc pre-service and in-service education offerings around sustainability. McCrea observed—among EC educators generally—*'shallow EE/EfS frames of mind'*, *'piecemeal ethical dispositions'*, and a *'lack of agency for inquiring/reflecting/provoking pedagogy'*. Such significant shortcomings can be addressed through EC teacher education, however, as Li states, *'pre-service and in-service programs without sustainability concepts'* are concerning.

The content of teacher education programs is typically informed by national ECE curricula. It is only relatively recently that some countries such as Australia, New Zealand, South Korea, Finland, and Sweden have incorporated sustainability concepts into their early childhood curricula (Davis & Elliott, 2024; Elliott et al., 2020). Both Ji and Mackey reflected on some of the challenges at the intersections between teacher education and sustainability curricula in their respective countries.

> In the revised 2019 Nuri Curriculum, it is stated, 'this curriculum is oriented towards a sustainable society'. Given this, many teachers want to know the relationship between the Nuri Curriculum and ESD. Therefore, along with clarification of some related concepts, I think there should be education for teachers on how they should implement the national level curriculum to become ESD. (OJ)

> Over the past 20 years of my teacher education career, EfS has undergone changes in response to world ecological disasters and events. [Nevertheless], responses have often lacked new

thinking and fresh responses, especially in teacher education. Some of the most significant research has shown the early years to be a time for creative solutions and problem solving (Davis, 2010). However, from personal observation, some responsibility lies within the thinking of curriculum developers who have not always understood that our youngest children have the curiosity and cognitive abilities to grapple with issues in our communities and on planet Earth. [Fortunately], recent curriculum developments in New Zealand, Australia and Scandinavia have embraced more culturally relevant and environmentally meaningful documents. (GM)

Again, the challenges cited speak to the ambiguities and misunderstandings around EfS concepts and foci. It is unsatisfactory to add a sustainability concept to ECE curricula with the expectation that this will be understood and embraced by all, from educators to teacher educators, assessors and professional development providers. Overall, *'new thinking and fresh responses'* are required across early childhood teacher education and curricula. Perhaps too, as McCrea proposes, a transformative shift from *'now orientations'* to *'intergenerational views about humans and other-than-humans across all layers of societies'* needs to be included in ECE.

Text Box 5.6

The catch phrase 'never waste a crisis; it's an opportunity' became prominent during the COVID-19 pandemic. Could you see these reported constraints as opportunities for change? If the listed constraints were addressed with effective resourcing, where could we be in another decade or two?

What change actions would you prioritise to address the constraints noted, so that every early childhood educator, teacher educator, early childhood service, national curriculum, and teacher education program promoted young children and their educators as active and vital participants in creating sustainable futures for all.

5.8 Emerging Forces and Pivotal Points in the Development of ECEfS

We invited our respondents to identify any pivotal points in ECEfS development over time and Wilson alerted us to her preferred term 'emerging forces'. Hence, we have adopted this term to capture comments from several respondents where forces seemed more apt. Wilson identified emerging forces as including, *'ideas about humans having a reciprocal relationship with the rest of nature, about children having the ability to be active participants in sustainability issues, and about appreciating the spiritual dimensions of the natural world'*. An emerging force for Inoue centred on issues of equity and social justice, precipitated by the UNCRC (UNICEF, 1989). She cited the consideration of children's active participation from infancy as transformative for images of children and suggested that *'even those who were not previously interested in social justice education in ECE have become interested'*. More broadly, she identified '*environmental education as an education related to a social justice worldview'*. Despite these positive forces, though, she also reminded us of the ambiguities around terminology and how *'the concepts of children's rights and participation are of Western origin, and it is necessary to consider how to deal with such concepts in non-Western cultures'*.

A strongly related emerging force is the increasing recognition of First Nations custodians of lands and waters, and the new field of giving rights to physical entities such as rivers or mountains. McCrea stated that '*people recognising first custodians more and a stronger sense of care for soil, water and air along with plants and animals'* as an emerging force. Ritchie offered a world-leading example from the Aotearoa/New Zealand early childhood curriculum, *Te Whāriki* (NZ Ministry of Education, 2017):

> In Aotearoa, our first national early childhood curriculum, Te Whāriki. He whāriki mātauranga mō ngā mokopuna o Aotearoa (NZ Ministry of Education, 1996) contained a strong commitment to Te Tiriti o Waitangi. Writers Margaret Carr and Helen May worked closely with Māori co-writers Tilly and Tamati Reedy which meant it included Māori language and knowledges which reflect a deep respect for Papatūānuku, Earth Mother and the environment. The 2017 revision expanded on this content to include a focus on kaitiakitanga [guardianship and protection] (NZ Ministry of Education, 2017). (JR)

Many fields of endeavour can point to inspirational leaders, advocates, and peak organisations that have consistently promoted their cause. In ECEfS, our respondents often returned to the international leadership of Professor Ingrid Pramling Samuelsson (Sweden) and co-author Professor Julie Davis (Australia) (see Chap. 2).

> Inspirational ECE leaders have moved EfS to the main agenda of early childhood discussions, such as the World ESD Awards initiated by Professor Ingrid Pramling-Samuelsson who also has the role of UNESCO Chair of Early Childhood Education. (GM)

> Professor Julie Davis, in collaboration with fellow Australian, Scandinavian and New Zealand researchers formed the basis for a thriving research hub that has led to widening the scope of doctoral research and international dialogue on EfS. (GM)

In addition, peak organisations such as OMEP and the TND research network were frequently recognised for their pivotal roles in scaling up ECEfS (see Chap. 2). Specifically, Engdahl cited '*The OMEP World Projects on ECEfS, collecting national contributions from between 20–40 countries, as reported and discussed at regional and world conferences'*; and Ji stated, '*I think the exchange of ideas, information, and the practices between the scholars from several countries through TND played an important role'*. As indicators of their success, a range of multiauthored publications have emanated from these organisations including several research texts, multiple papers, and various practitioner publications (see Chaps. 2 and 4). Pramling Samuelsson noted that '*publications are expanding on the topic of ESD and young children, although there are many black spots'*, and some countries are yet to engage with this topic. We look forward to seeing these gaps filled in the not-too-distant future.

On a national scale, the Australian Association for Environmental Education Early Childhood Special Interest Group was perceived as *'significant in developing a national presence' (MH)*, along with the Australian state-based networks (see Chap. 2). Both Hocking and McCrea highlighted recent Australian policy shifts including the *National Quality Standard* (ACECQA, 2018) and the revised *Early Years Learning Framework* (AGDE, 2022) as pivotal for further EfS uptake (see Chap. 2). Nevertheless, as McCrea reminds us, any important policy shifts must be guided by a '*more overarching vision with a deeper ECEfS ethos and be ongoing across whole setting practices*'. In addition, she suggests *'there are questions of meaning - what are reasonable quality and quantity parameters/examples of 'sustainability' for each setting, for various organisations and governments?'* Such questions may inform future turning points in ECEfS, as we recognise that there will be many such turning points as newer and, as yet unrecognised forces emerge into the future.

Text Box 5.7

Irrespective of your own journey in ECE and sustainability, what do you perceive as emerging forces or pivotal turning points? Consider what broader forces might promote systemic uptake of ECEfS. Where/who are the pivotal people, organisations, publications, and other resources that are standing out and making a difference across the field?

5.9 Thinking About Future Trajectories

Lastly, respondents offered both hope and explicit guidance about future trajectories for ECEfS, informed by the dual urgencies of addressing climate change and recognising the potential of EfS as integral to high-quality ECE. Ji, Ritchie, and Pramling Samuelsson expressed these complexities from their different socio-cultural and geographic positions.

> The climate crisis will become more serious in the future, and since 'education' is a key factor in SDGs, governments and scholars of all countries will pay more attention and make various efforts. On the other hand, I believe that inappropriate attempts to teach some concepts related SD will appear... The efforts of experts like us should be strengthened to ensure that ECEfS is properly implemented. (OJ)

> Currently Aotearoa, like many parts of the globe, is experiencing severe impacts of climate change. It is my hope that in reaction to the existential peril that is being dramatically highlighted, our government will prioritise funding for professional growth opportunities for educators to equip them to better support children and families in these increasingly scary times. It is still a struggle to convince our Ministry of Education to prioritise quality early years education even though we know that this is the period of life in which children form important dispositions such as empathy and caring for one another and the environment. (JR)

> I think ESD has to be an evident and decisive aspect of quality in ECE, since now there is a tendency that quality work is one thing, and ESD is something else. I want to see more empirical work, as many publications in this field are more theory or ideology driven, than based in empirical data. (IPS)

Another way of considering this issue is to think of ECEfS more as a social movement, than a curriculum matter. As Engdahl argues, *'Sustainability is not a subject; it is a concept for a movement that collects multiple initiatives and actions for the complex and most important issues of our time. There isn't an alternative NOT to engage'.*

Consequently, how do we navigate and engage with these urgent and intertwined complexities? Mackey reiterated young children's roles as citizens with rights and sees *'a growing awareness of the meaning of citizenship, with questions being asked as to how this positions our youngest citizens'*. Across the Pacific, Wilson also cited a *'greater understanding and appreciation of children's agency'*. In Sweden, Engdahl argues *'we need to always give voice, space, and respect for young children, to promote their perspectives and ideas'*. Pramling Samuelsson boldly proposed and continues to lobby for, a *'UN DECADE for ECE that would help to put the youngest children on the agenda'*. There is no question amongst our respondents that children's voices are fundamental to all future ECEfS trajectories.

There is also an argument for widening the field to embrace further perspectives, actions and ways of engaging with EfS. Ji suggests we seek diverse ways *'to inform the public of the importance of ECEfS through influential cultural and artistic activists'*. She offered the example of a Korean boyband that performed at the SDGs 'Moment' in Korea's 76th General Assembly 2021, leading to worldwide attention. In another approach to widening the field, Li described ECEfS as both *'a global and local journey'* reiterating that *'the concept of ECEfS needs to be adapted to different cultural contexts'*. Thus, we return to the idea of networking as so fundamental, with Engdahl advocating for us *'to network with all kinds of associations, from Fridays for Future to NGOs, and with politicians, for transformative and transactional actions towards a sustainable future'* to widen the audience. Alternatively, Wilson suggests embracing all—including the human and other-than-human—within an ecological community:

> The field will grow wider by engaging with more diverse groups and by expanding the idea of "the ecological self" to "the ecological community". Our understanding of "ecological identity" will also encompass a "social justice" component, focusing on not only more marginalised human groups, but also on other-than-human entities.

Finally, Inoue makes a strong case for collaborative research with educators and children as instrumental to any future trajectories. She states:

> In ECEE research, it is not enough to simply present theories; it is necessary to conduct ongoing collaborative research with the field of practice to reflect these theories in the practice of real ECE services. It is not enough to simply have playgrounds which are rich in nature; it is necessary to examine how children's views of nature are transformed in that playground, how this transforms their views that lead to nature conservation, and how it changes their lives, and to present a model for practice.

Our respondents demonstrated that there are many ways to guide future trajectories in ECEfS, with many more likely to be proposed beyond what this text offers. Hocking positively envisions that *'the trajectory of ECEfS should travel onwards and upwards',* while both Inoue and McCrea acknowledge the pressing challenges but offer hope.

> The reality of the natural environment is worse and the social situation is so chaotic now. Sometimes I feel there is no hope. But, we have to move forward with hope. (MI)
>
> I believe there is still a long journey ahead as there are sustainability struggles in terms of values, meanings and practices. But hopefully, we are on a promising path towards a passionate everydayness orientation with ECEfS, both within education systems and services plus with educators and families joyfully enacting wholistic sustainability. (NM)

Text Box 5.8

Respondent McCrea states, *what I wish for going forward is—everyone sensitively enacting deeper ways of understanding, caring, sharing and engaging with/in all dimensions of sustainability.* What do you wish for, going forward?

Our respondents have shared many ways of promoting and engaging with future ECEfS trajectories. Which ones resonate with you? What other ideas can you offer? Perhaps, more critically, how can you swiftly enact these ideas and what do you see is your own pivotal role?

5.10 Concluding Thoughts

As we discussed in this chapter's introduction, we wished to canvas the views and ideas of a number of our international colleagues who have been on our shared journey over the years in building momentum for ECEfS. It is generally agreed that, initially at least, many of us were individuals working away in isolation within our geographical spheres and spheres of influence. However, as the comments embedded here reveal, we now truly have an international movement that has gone from strength to strength over the past four decades. We believe this has occurred because we have been able to tap into individual concerns for the state of the planet and humanity and especially for young children's future prospects. More importantly, however, we have all witnessed that liaising, advocating, collaborating, sharing, and learning together have been the essential elements leading to the international development of ECEfS. No one has truly led the field; we have all relied on each other, along with many contributors unknown to us. We offer our thanks and appreciation to all who have shared this unique and reflective journey at a critical time for children and the Earth.

Appendix: Interview Questions Shared with Participants for Written Responses

1. What was the initial trigger for your involvement in early childhood EE/EfS/ESD?
2. Who or what influenced you most?
3. What do you see as your personal contribution to early childhood EE/EfS/ESD?
4. What maintains your involvement, if any, now?
5. Over time, what do you identify as key enablers and constraints in the evolution of this field?
6. What do you consider were pivotal turning points in the development of early childhood EE/EfS/ESD?
7. What do you see as the future trajectory of this field? Why?
8. Any other thoughts you may wish to share?

References

Australian Children's Education and Care Quality Authority (ACECQA). (2018). *Guide to the National Standard*. ACECQA.

AGDE. (2022). *Belonging, being and becoming: Early years learning framework for Australia (V2.0)*. Australian Government Department of Education for the Ministerial Council.

Carson, R. (1956 republished 1998). *The sense of wonder.* Harper & Row.

Carson, R. (1962). *Silent spring.* Houghton Mifflin Company.

Carter, F. (1976). *The education of little tree.* Delacorte Press.

Davis, J. (2010). What is early childhood education for sustainability? I J. Davis (Ed.), *Young children and the environment early education for sustainability* (pp. 21–42). Cambridge University Press.

Davis, J., & Elliott, S. (2024). *Young children and the environment: Early education for sustainability* (3rd ed.). Cambridge University Press.

Davis, J. M., & Davis, J. E. (2020). Early childhood teacher education and education for sustainability: A review of the literature and mapping of courses. In S. Elliott, E. Ärlemalm-Hagsér, & J. Davis (Eds.), *Researching early childhood education for sustainability: Challenging assumptions and orthodoxies* (Vol. 2, pp. 111–124). Routledge.

Davis, J., & S. Elliott (Eds.) (2014). *Research in early childhood education for sustainability: International perspectives and provocations*. Routledge.

De Chardin, T. (1957 republished 2004). *The divine milieu.* Sussex Academic Press.

Elliott, S., Ärlemalm-Hagsér, E., & Davis, J. (Eds.) (2020). *Researching early childhood education for sustainability: Challenging assumptions and orthodoxies* (Vol. 2). Routledge.

Engdahl, I. (2015). Early childhood education for sustainability: The OMEP world project. *International Journal of Early Childhood, 47*, 347–366. https://doi.org/10.1007/s13158-015-0149-6

Evans, N., Harwood, D., Furu, A.-C., & Alici, Ş. (2022). Challenges in reorienting early childhood education for sustainability in initial teacher education: Transforming, creating and expanding. *Australian Journal of Teacher Education, 47*(5). https://doi.org/10.14221/ajte.2022v47n5.1

McCrea, N. (2015). *Leading and managing early childhood settings: Inspiring people, places and practices.* Cambridge University Press.

Mackey G., & Ritchie J. (2012). *Pedagogies for Sustainabilities in Early Childhood Education in Aotearoa New Zealand*. York University, Toronto, Canada: Seventh Biennial Meeting of the International Network of Teacher Education Institutions UNESCO, May 16–20, 2012.

New Zealand Ministry of Education. (2017). *Te Whāriki*. New Zealand Ministry of Education.

Pramling Samuelsson, I., & Kaga, Y. (Eds.). (2008). *The contribution of early childhood education to a sustainable society*. UNESCO. http://unesdoc.unesco.org/images/0015/001593/159355e.pdf

Pramling Samuelsson, I., Li, M., & Hu, A. (2019). Building a sustainable future for our young children in China and Nordic countries. *ECNU Review of Education, 2*(4), 369–373. https://journals.sagepub.com/toc/roe/2/4.

Ritchie, J., Duhn, I., Rau, C., & Craw, J. (2010). *Titiro whakamuri, hoki whakamua. We are the future, the present and the past: Caring for self, others and the environment in early years' teaching and learning*. Final Report for the Teaching and Learning Research Initiative. http://www.tlri.org.nz/tlri-research/research-completed/ece-sector/titiro-whakamuri-hoki-whakamua-we-are-future-present-and

Siraj-Blatchford, J., Pramling Samuelsson, I, Lenglet, F., et.al., (2010). *Taking children seriously–How the EU can invest in early childhood education for a sustainable future*. European Panel of Sustainable Development, EPSD. Report No. 4, 2010-12-17. GMV, Centre for Environment and Sustainability: Chalmers University of Technology. www.ecesustainability.org

Siraj-Blatchford, J., Smith, K. C., & Pramling Samuelsson, I. (2010). *Education for sustainable development in the early years*. OMEP Sweden. https://files.builder.misssite.com/9c/75/9c7543cb-1380-45f9-9440-e7a16a1e42c6.pdf

Chapter 6
Key Learnings, Challenges, and What Next

The future is not some place we are going to but one we are creating.
The paths to it are not found but made,
and the activity of making them
changes both the maker and the destination.
(*Australian Commission for the Future*, 1989).

Abstract This final chapter summarises our key learnings about, areas for amplification, and remaining challenges for ECEfS. As a field in evolution, we understand that what seems relevant and urgent now may be prioritised differently in coming decades, especially—we hope—as new practitioners, researchers, and advocates emerge who challenge current Western-centric views. We note with some satisfaction that there are national and international policy shifts in ECE that are further advancing ECEfS into a global movement. The story of ECEfS is far from finished!

6.1 Introduction

The other side of history is the future. A key purpose of this slim volume about ECEfS history is to further mobilise the ECE field to understand that the crisis of sustainability, especially the climate crisis, is a children's crisis. Given that EC is the most critical phase of development in the human life span, we reflect that ECE must prioritise EfS/ESD as foundational for all children's early learning. This is vital to address the legacies of past omissions and slowness to act on climate change and sustainability and urgently address present and future challenges for children and future generations. We argue ECEfS should simply be the norm in all settings where young children are concerned. While ECEfS has become a serious field of educational endeavour over the past few decades, we are still at the beginning of this journey to bring everyone, everywhere, on board. Although we have been strongly advocating for ECEfS for over 30 years, it is beyond time to make the shift from advocacy to engagement, implementation, and action!

J. M. Davis et al., *Early Childhood Education for Sustainability*,
SpringerBriefs in Education, https://doi.org/10.1007/978-981-97-8405-9_6

Nevertheless, in this short history, we claim some strong achievements. We see a maturing field that is moving from its initial focus on practical strategies—as important as these are—towards broader theorising that offers ethical and intellectual frameworks for organising and helping overcome the often idiosyncratic and shallow approaches to implementation. In particular, shifting from benign forms of EE focused on nature play to greater political understandings of the power of early education to contribute towards shaping children's worlds now and into the future, is one achievement, even if there remain pockets of resistance to such ideas. Allied with this, we also see the rapid expansion of an ECEfS research culture guiding and challenging thinking and practice as contributing significantly to such achievements.

Further, the adoption of new or revised ECE curriculum frameworks in a growing number of countries where sustainability—in its broader, multidimensional form—has become a core curriculum feature is also to be celebrated. Notable curriculum exemplars include New Zealand, Norway, Sweden, Finland, Korea, and in 2022, Australia. None of these achievements would have occurred without strong and persistent arguments and advocacy by individuals leveraging their influence within and across multiple local, national, and international networks.

6.2 Areas for Amplification

In this final chapter, we outline our ideas for key ECEfS areas requiring further amplification as we move forward but note these are likely to present tensions and challenges for the field; this is not the time for taking our feet off the peddle! Here, we emphasise seven areas of shifting and overlapping discourses and discussions that offer further growth for ECEfS, if taken on board. These are: enacting systems change vs individual change for sustainability; rethinking nature-human relationalities; shifting images of children; expanding children's rights and justice issues; promoting children's health and well-being; expanding diversity of constructs and cultural contexts; and capturing the wave of global trends for socio-political change for sustainability.

6.2.1 Systems Change Versus Small-Scale, Individual Changes for Sustainability

Our long experience and ongoing research tell us there remains strong focus on the power of individuals to enact change for sustainability. While we do not wish to diminish individual efforts, there is overwhelming evidence that it is only through working with systemic, multistakeholder change models will we realise wider and deeper professional, educational, and community engagement and transformation for sustainability. Achieving sustainability is both a personal and a collective project.

We must go well beyond 'excellent champions', except where tied intentionally to outreach and wider learning for change within organisations and systems.

The individual behavioural change narrative is powerful in society and education and remains powerful in EfS (Ferreira et al., 2021). Inevitably it is powerful in ECE although there are counter-narratives such as inclusion, community, social justice, and democracy. While ECE often highlights efforts in creating citizens to live in democracies, it also sits within educational ecosystems that promise to advance individualism and individual success. There are more efforts about reducing waste, water and conserve electricity within individual classrooms and centres than there are about transforming ECE systems and communities. It is well-recognised that there are limits to individual behavioural change in terms of improving environmental outcomes to scale (Kenis & Mathijs, 2012), but we also recognise the structural impediments that operate to maintain business as usual while 'allowing' small-scale individual change efforts to be valorised, provided they do not over-emphasise breaking down existing structures and relationships. For example, current awards programs in Australia and perhaps elsewhere tend to celebrate and recognise individual services and the work of individual educators in EfS, but they labour under inadequate sustainability-enhancing systems, be they curricula, policies, governance, procurement, and management protocols. These can be viewed as giving tokenistic appreciation and appeasement to sustainability advocates, demonstrating some support without shifting EfS to systemic mainstream.

Nevertheless, there are exemplars where collective and social learning for sustainability predominate, particularly from non-Western countries (Almeida et al., 2024). Recently collated stories from the Australian field (Davis & Elliott, 2024) demonstrate how some centres *are* fully embracing ECEfS in systemic ways, recognising the multidimensionality of sustainability with children and, at the same time, transforming their communities. We believe, and research backs this up, that systemic, multistakeholder change approaches are vital across the diversity of ECE services and settings to drive wider, educational, professional, and community engagement and transformation. Innovative partnerships and alliances could be strengthened and supported through, for example, developing an international accreditation framework such as Kagawa's *Climate Resilient Education Services* (2022) sponsored by UNICEF to underpin sustainable infrastructure, climate mitigation strategies, and transformative curricula. This strategy offers systems-wide standards and progress indicators for climate resilient education systems that could be tailored towards ECE services in Australia and elsewhere. We see such efforts as vital for reaching deeper and wider into and across the ECE sector.

6.2.2 Rethinking Nature/Human Relationalities

Plumwood (2002) was an early voice talking about human's ecological embeddedness and vulnerability. She argued for cultural change to 'exercise responsibility and care for other and for the natural world' (Plumwood, 2002, p. 240). Global biodiversity loss is a stark reminder of how humans have failed in their care for the natural world. Since Plumwood's initial call for change, several theoretical shifts have developed. These include new ethical relationalities with worldviews incorporating understandings of human and non-human existing and inhabiting the world together—as opposed to a binary understanding. The concepts of the more-than-human (Haraway, 2016) and common worlds inform the post-human theoretical research field (Taylor, 2013, 2016, 2017).

Post-humanism, according to Taylor (2016), includes a diverse spectrum of studies including, for example, animal studies, 'new' material feminism, assemblage theories, actor network theory, and decolonial and indigenous studies. In ECE studies, such theoretical perspectives are being developed by researchers (see Chaps. 3 and 4). However, while such theoretical rethinking is yet to significantly filter through to practitioners, we note that it has given impetus to the resurgence of nature play programs internationally (Hughes, 2021).

Additionally, the rights of nature have emerged as a theoretical discussion point. As noted earlier, Anthropocentrism translates into the belief that human well-being is our central consideration. This orientation renders the non-human world open to over-exploitation of land, waters, the atmosphere, ecosystems, biodiversity and species, and alarmingly, inner and outer space, for the 'good' of humans. To overcome this destructive binary, some jurisdictions, for example New Zealand, has granted the rights of personhood to the Whanganui River, recognised as a living entity with its own identity and legal protections within human law, drawing on Māori understandings of the river as an ancestor—a tupuna -a 'person' with the river's well-being directly linked to that of humans. More generally, such interconnections between the human and non-human can be found in other First Nations societies, including the Australian Aboriginal and Torres Straits Islander peoples and the Sami of Scandinavia, where customary law and internalising of ecocentric principles interpret the natural world as sacred and, therefore, as a tool for maintaining social and ecological balance (Wong, 2019). There is much here to be valued and amplified to potentially disrupt predominant Western binary nature/human relationalities.

6.2.3 *Shifting Images of Children*

Contemporary thinking about young children, curriculum, and early childhood learning and teaching must inform this reorientation. In Chap. 2, we flagged that shifts are underway across policy and practice; however, it must be recognised that across diverse cultural contexts, there is no one image. Culturally relevant ways must be sought to empower all children to act sustainably and become eco-resilient who can deal with climate change trauma and challenge the systems around them. Capacities to critically examine, question, problem-solve, work collaboratively, act intentionally, and engage in ethically informed decision-making are core. The UNESCO (2017) competency framework offers further guidance with eight key competencies: systems thinking, anticipatory, normative, strategic, collaboration, critical thinking, self-awareness, and integrated problem solving. For educators to pedagogically scaffold such competencies in ECE, informed by contemporary early childhood curricula that embed sustainability principles (AGDE, 2022; NZ Ministry of Education, 2017; Swedish National Agency for Education, 2019), these competencies offer a firm foundation for amplification of children's actual and perceived roles in their communities.

6.2.4 *Expanding Children's Rights and Justice Issues*

Intersecting with discourses about nature/human relationalities and images of children are those about rights and justice. This focus for continued amplification concerns giving much greater attention to children's rights in the context of sustainability and climate disruptions, and recognition of the limits when focusing simply on human rights. To amplify the first point, we observe that far greater attention is given to children's rights in the Anthropocene by lawyers and medical/health professionals than by EC educators, even though EC educators are reputably more 'in touch' with children and young people than most lawyers and health specialists. For example, Fenton-Glynn (2019) and chapter authors in her edited text *Children's Rights and Sustainable Development: Interpreting the UNCRC for Future Generations* clearly recognise that children fare worse when communities face social, economic, and environmental challenges and, at the same time, lack both private and public voice in such unjust matters. Aligned with these advocates for child rights-based approaches, we call for ECE and ECEfS to more strongly recognise and emphasise the grossly unfair impacts of unsustainability on children and those not yet born.

We recognise, too, that such impacts on children are felt disproportionally across the globe. The 2022 COP 27 fund to shift some resources from the global north to the global south/majority world countries to manage climate change impacts is some recognition that the global north has a responsibility to all children, not just those in their own country. As noted in Chap. 2, to rethink children's rights in the Anthropocene, Davis' (2014) expanded rights framework might be useful. While continuing to emphasis the centrality of children's rights, this offers a multidimensional framework of rights, both human and non-human into the future. Additionally, the recent UN initiative of the Committee on the Rights of the Child (CRC) (2023) further emphasises children's rights with a special focus on climate change.

6.2.5 *Promoting Children's Health and Well-Being*

Intertwined with reconsideration of rights are growing concerns about the health impacts of climate change for young children. Access to clean water, food, and shelter are impeded for many children globally and specifically, amid climatic events. Such impacts may derail lifelong developmental trajectories with a lasting burden on the social, cultural, and economic dimensions of sustainability. Attention has been drawn to this by recent Lancet publications (Bansal et al., 2023; Clark et al., 2020; Helldén et al., 2021). Bansal et al. (2023) argue the necessity for longitudinal studies about climate impacts from in utero to better inform public health responses, potentially mitigating lifelong negative consequences. More broadly, the terms eco-anxiety and solastalgia (Farrant et al., 2012), are increasingly employed to describe the reported emotional distress and despair of youth about their futures and the planet (Baker et al., 2021; Sanson et al., 2019). Cooke and Baldwin (2024) recognise synergies between the principles of health promoting early learning services and whole centre EfS approaches with potential to be mutually reinforcing. Ramping up understandings of links between climate change and children's health, both short and long term, is essential.

6.2.6 Expanding Diversity of Key Constructs and Cultural Contexts

We recognise that sustainability, and ECEfS for that matter, has been largely championed by Western nations and Western-orientated researchers and advocates, like ourselves. Indeed, it can be interpreted as a Western intrusion akin to colonialism (Brinkmann, 2022) that perpetuates normative (Western) ideas and ideals about children, early education, and sustainability (Arts, 2019). In acknowledging such viewpoints, and to counter these arguments, we recognise that there must be a diversity of culturally informed and culturally inclusive ECE approaches that emerge from the diversity of contexts across regions continents and countries (Almeida et al., 2024; Dreamson & Kim, 2022; Inoue, 2014; Ritchie, 2014). These must be built on the needs, aspirations, and actions of local people and communities; in many cases, first focus should be on improving basis standards of living–housing, sanitation, food, and water security—rather than 'loftier' ideals such as green energy, reducing CO_2 emissions, or enhanced waste recycling.

Central to rethinking and repositioning taken-for-granted viewpoints, ECE must strengthen recognition of worldviews and knowledges of First Nations Peoples across the globe and their unique contributions to sustainability over thousands of years. For example, in the Scandinavian context, children belonging to Indigenous Sámi groups should be supported in developing cultural identities and local language learning (Swedish National Agency for Education, 2019). Laiti and Määttä (2022) argue that ECE has an important role in maintaining, transferring and strengthening Sámi languages and cultures, key aspects of cultural sustainability. Their study demonstrated that educators can actively develop Sami culture amongst children, especially 'the values, world view, and practices of the Sámi culture' (p. 72) as 'Sámi culture and language are adjusted to the local environment, location, time, rhythm of nature' (p. 73). In Jannok-Nutti's (2010) thesis emphasising Sámi culture and

Sámi traditional knowledge, mathematics in preschool and school was explored through traditional measurement units. A Finnish study (Laiti et al., 2022) illustrated how Sámi culture and identity could be strengthened through everyday activities and interactions where Sámi languages in storytelling and music strengthened Sámi identity. These authors stressed that—as nature is a focal aspect of Sámi worldviews and an essential nature-human connection in maintaining the sustainability of the Arctic—nature learning must be fundamental to understanding Sámi culture:

> Sámi early educators' main goals for Sámi early childhood education were learning the language, supporting children's Sámi language learning, supporting children's identity formation, positive attitude towards Sámi culture and language, and cooperation with Sámi community. These were implemented and connected with the help of Sámi content connected with nature, authentic livelihoods, or traditional handcraft, and using Sámi language. Nature was deemed the most valuable basis and resource … and … children's identity formation was supported by emphasizing the connection to nature and usage of Sámi language. (Laiti et al., 2022, p. 793)

6.2.7 Catching the Wave of Global Trends for Socio-Political Change for Sustainability

As we move into the next phases of ECEfS, we are beginning to see the amplification of global trends that have the potential to strengthen the field and move it beyond where we are currently. For example, increasing inequities, regional conflicts and wars, the rise of populist, anti-democratic politics, increases in misinformation and disinformation, and the COVID-19 pandemic have all been drivers for civic unrest and calls for social change. Of particular concern to us are the compounding climatic disruptions coming at a quicker pace and potentially overtaking forecast global temperature tipping points (IPCC, 2023). With dire consequences for humanity—particularly for young people and future generations—and for the Earth, perhaps this is a timely confluence of issues to dramatically rethink and shift the ways we live and interact across all species.

While there are some grave challenges associated with current trends, we have also seen the emergence of major progressive social movements, often global in scale, such as #blacklivesmatter, #school strikes for climate, and #metoo. Perhaps a renewed global peace movement—similar to that of the 1960s and 1970s at the height of the cold war—will appear as events unfold in the Ukraine war in Europe, and USA/China posturing with accelerating tensions in the Asia–Pacific.

As our own brief profiles in the Preface illustrate, our interest in ECEfS was itself a response to concerning global issues emerging in the last century. We believe that ECEfS is responding to and helping to shape thoughts and actions in this contemporary milieu of growing social, economic, and environmental challenges and possibilities. Expanding ECEfS globally and further amplifying our messages for a global green and equitable transition to sustainability remains our key task.

6.3 Identifying and Overcoming Challenges: Some Exemplary Strategies and Approaches

At the beginning of this chapter, we discussed conceptual and theoretical areas within ECEfS that we see as requiring further amplification if the field is to boldly march forward. In this section, we focus on the broader socio-political and policy context, then explore ECE-related challenges we have identified as working to maintain the status quo and against our hopes for expanding ECEfS. While these hurdles may, at times, seem overwhelming, we also offer exemplary strategies and approaches that point towards overcoming these challenges, based on our ongoing international research and our anecdotal evidence from deep involvement in this field now for decades.

But first, there continues to be significant inertia, and in some case outright resistance, in the broader political landscape to considering topics and issues connected with the substance of this book. The 'climate wars', 'culture wars', and the rise of far-right politics, disincentivise action around sustainability because a lack of political will around this topic has potential for alienating some loud and reactive sections of the population. Such a potentially toxic social-cultural milieu has made achieving traction for sustainability efforts quite difficult, especially in Australia. Nevertheless, despite ineffective political leadership around sustainability in many countries, civil society movements, social media platforms, and some business groups are pushing for 'greener' policies, while more progressive political parties and independent political voices are helping to push in the right direction. Nonetheless, political change *is* possible and, in Australia in May 2022, a change of government shifted the balance towards more climate and sustainability-friendly politics and policies. Examples include an expanding push towards electric vehicles, policies to reduce single-use plastics, increases in renewable energy production, and better biodiversity outcomes.

Internationally, successive and compelling IPCC (2023) reports, UNESCO initiatives like the SDGs (UN, 2015) and the UN Conference of Parties (CoP) meetings, alongside progressive social movements, suggest momentum for change is building. However, intersecting sustainability with ECE and its accompanying raft of gender, workforce, and access issues offers a political hotspot for many governments internationally. It is against this somewhat heated and fluctuating socio-political context that we prioritise five ECE-related challenges and offer some exemplary strategies and approaches to meet these.

6.3.1 Challenge 1: Overcoming Curriculum Policy Inertia

Allied with governmental inertia in many countries has been the lack of national curricula for ECE that embed sustainability. Reorienting and updating early childhood curricula to put sustainability at the core are necessary and concrete policy actions to ensure children's active participation in climate education and EfS, with understandings of global sustainability being central to ECE programs and practices. Only a small handful of countries including Sweden, Norway, Finland, New Zealand, and South Korea have sustainability as a core curriculum principle. In 2023, we were delighted to add Australia to this list—an achievement 25 years or more in the making!

While enshrining sustainability into curriculum policy is a crucial first step for embedding systems-wide change, this must be followed by effective and well-resourced teacher education and professional learning to promote uptake with ongoing evaluative research essential to this process. Some illustrative strategies and approaches are offered to overcome curriculum policy inertia.

A. One example of a systems approach to curriculum/pedagogical change came from actions instigated by participating member states at the 2022 CoP26 in Glasgow. Prior to *The Joint Event of Education and Environment Ministers Summit—Together for Tomorrow: Education and Climate Action'* (hosted by the United Kingdom Department for Education, Italy, UNESCO, and Youth4Climate), Ministers responsible for Education, and/or Climate Change, and others with responsibilities for young people from across the globe, discussed the importance of education in creating climate positive futures. Commitments led to pledges from many countries to embed transformative EfS into their education systems with repeated reference to EfS needing to start in ECE. They also championed the necessity for teacher education around climate change, climate action, and sustainability.
B. In 2022, the European Union (EU) released their *Proposal for a Council*[1] *Recommendation on learning for environmental sustainability* that sets out recommendations for member states to urgently step up efforts to support education and training systems for Europe's green transition, so that all learners from all backgrounds could access high-quality, inclusive education and training on climate change, biodiversity and sustainability. This Proposal clearly emphasises for ECE to play a key role as the starting point for ongoing EE/EfS. The accompanying *Staff Working Document on Environmental Education for Sustainability* details the concepts and recommendations put forward in the *Recommendation*, with research evidence and good practice from across Europe presented as a guide for policymakers, educators, individuals, and all organisations working on sustainability in the EU education and training sectors. This Staff Handbook involved wide public consultation including with early childhood educators in Europe, as well as Davis and Elliott, emphasising the growing recognition of ECE and its contributions to EE/EfS.
C. A final example of synergies that are evolving to put systems approaches into ECEfS are the innovative partnerships and alliances across the ECE system in the UK through the international accreditation framework—*Climate Resilient Early Childhood Education Services* (CRECES) (akin to Ecoschools)—to underpin

[1] European Council comprises the heads of state/government of the 27 EU member states, the European Council President, and President of the European Commission.

sustainable infrastructure, climate mitigation strategies, and transformative curricula.

6.3.2 Challenge 2: Overcoming Resistance to Pedagogical Change for Sustainability

Educators' pedagogical repertoires are informed by many sources including their underpinning images of children, values and worldviews, education and training, theoretical and research backgrounds, and the curriculum policies with which they work. Finding ways to facilitate pedagogical change requires multiple avenues, particularly when dealing with potentially contentious issues such as climate change. However, there are many aspects of ECE pedagogy that clearly align with implementing EfS and have been well-identified as advantageous (Elliott & Davis, 2009; Pramling Samuelsson & Kaga, 2008). In addition, Chap. 3 offers theoretical and pedagogical foundations for exploring new ways of embracing pedagogies for sustainability and bridging the policy guidance gap often cited around ECEfS (Weldemariam et al., 2017).

Finding pragmatic leverage points to overcoming resistance and begin change processes are critical - it might be educator interest in nature play, children's health concerns, or a passion for creative upcycling. Cascades of change are feasible once educators realise the potential of working closely with children. Opportunities to use action-oriented inquiry or project approaches (Elliott & Pugh, 2020; Ji & Stuhmcke, 2014), to explore transformative pedagogies (Mezirow, 1997), slow pedagogies (Clarke, 2023), place-based pedagogies (Hughes et al., 2022), trauma-informed pedagogies (AGDE, 2022) or environmental play-based approaches (Cutter-Mackenzie et al., 2014) offer much scope for pedagogical change in ECE. We acknowledge that, almost universally, workforce issues and the legacy of COVID concerns are currently impacting the ECE field, potentially thwarting the pedagogical changes urgently needed to address sustainability. Therefore, some guiding exemplars for change are shared below.

A. In an attempt to bridge the policy-practice gap, Elliott and Davis (2023) offer six pedagogical principles: collective action for change; transformative pedagogies; ethically informed relationships; democratic and civic participation; awareness of worldviews, values and cultural wisdoms; and the promotion of children as agentic with extensive rights. All have implications for daily pedagogical interactions with children and could scaffold educator transitions from policy to practice.
B. The emerging global nature play movement—see Chap. 3—offers a pedagogical provocation or leverage point for change. Regular engagement with children in local environments can bring local sustainability issues or concerns to the fore for examination and action with children. This may involve seeking specialist expertise from park rangers, Indigenous elders, biologists or wildlife carers, thus

broadening community partnerships for sustainability. Illustrative examples are available in a range of literature including Hughes et al. (2022).

C. Professional interest in developing theoretical and pedagogical strategies offers opportunities for enhancing EfS in ECE. In Sweden, 300 municipality leaders, principals, preschool pedagogical development leaders, and preschool teachers representing around 200 preschools from nine municipalities/providers, took part in a research and development program (2021–2024) entitled *Sustainable Preschool* (Ifous, 2021). In the first phase, researchers and participants collaborated in developing in-depth knowledge and competencies with the teachers, principals, and operations managers to organise and lead EfS. Findings identified that the teachers were interested in EfS, with some longing for more knowledge about ECEfS and transformative change. However, they also needed courage and professionalism in leading to find relevant EfS content and activities. The study revealed that teaching for sustainability in preschool requires developing knowledge, creativity, problem-solving skills, critical thinking, action skills, innovative thinking, and a change orientation. Further, it was recognised by participants that children's participation for a sustainable present and into the future is a decisive element if change is to occur (Ärlemalm-Hagsér et al., 2023). In 2023, the program focused on leadership and how participants can implement a *Whole School Approach* into management and organisational strategies.

6.3.3 Challenge 3: Scaling up Early Childhood Teacher Education

Here, we discuss initial early childhood teacher education (ITE) and vocational training, and the professional learning (PL) of teachers already in the field.

6.3.3.1 ITE and Vocational Training

One of the glaring 'holes' in creating systemic change for ECEfS lies with teacher education. Reorienting preservice education, course accreditation, and professional standards are essentials for early childhood educators and tertiary educators to prioritise sustainability and climate education. While many universities that prepare EC teachers have stepped up in relation to their high-level strategies—such as responding to the SDGs—and are expanding sustainability research efforts, research has shown that such efforts do not always filter down to specific ECE faculties, programs, and courses (Davis & Davis, 2020; Evans et al., 2022). We consider it imperative that ITE and training authorities such as vocational institutes and their regulatory bodies take serious responsibility to 'teach the teachers' so they can implement quality ECEfS with children. EC teacher educators must be leaders through their own course and subject offerings and research efforts aimed at providing beginning educators with the ideas, dispositions, skills, and resources to be 'agents of change' within ECE.

New areas we identify for prioritising include: transdisciplinary/interdisciplinary curriculum in ITE; environmental justice/intergenerational justice; leadership for sustainability; disaster resilience education and trauma-informed pedagogies; and action research approaches to learning and assessment with a sustainability focus.

Examples related to the above areas are scant; however, some exemplars are offered here.

A. In Türkiye, Şule Alici conducted a doctoral study highlighting a multiplicity of factors influential in the uptake of ECESD in ITE. Shifting government/institutional policies, reviewing ITE program structure/content, and upskilling teacher educators were perceived as relevant starting points to build ESD understandings and practice (Alici, 2018, 2020). Alici advocates for a competency-based framework and systems approach to facilitating change in ITE and overcoming pervasive ad hoc approaches. She also suggests there is scope for more innovative pedagogies including media literacy. Turkish researchers Akyol et al. (2018) employed drama studies with pre-service teachers to engage in 'hands-on and minds-on' approaches to ESD content.
B. Glynne Mackey, from Aotearoa/New Zealand (Mackey & Ritchie, 2012), had a long history advocating for EfS within ITE at national and international levels. The decade-long UNESCO focus on reorientating teacher education towards ESD led to a plethora of online resources available for any institutions seeking to address ESD in ITE programs. Another avenue for change demonstrated by Mackey was to facilitate the inclusion of sustainability and social justice in an Aotearoa/New Zealand national review of the teacher code of ethics and standards for all registered teachers. When such benchmarks are nationally required for all registered teachers, it is contingent on ITE institutions to comply by incorporating relevant EfS content within their programs.
C. Recent studies addressing the gap in ITE EfS research offer insights into avenues for facilitating change (Evans et al., 2022; Tilbury & Mulà, 2023). The nationally based case studies collated by Tilbury and Mula (2023) highlight many key strategies, ranging from teacher educator networking to experiential pedagogies and ESD policy frameworks. The lists are instructive and suggest a multiplicity of approaches are required, as no one strategy will suit all contexts. In the special journal issue (Evans et al., 2022), research by Spiteri (2021) and Ärlemalm-Hagsér et al. (2022) argues for exploring pre-service teacher perceptions of ESD, while Berry (2022) and Weldemarian (2022) introduce the potential for post-humanist discourses, with the intention to deepen the thinking of pre-service teachers beyond practice-based initiatives.

6.3.3.2 Professional Learning (PL) for Educators in the Field

Delivering PL for educators around sustainability for those already in the field is vital to overcoming resistance and fear of engaging with ideas and practices that may make some practitioners uncomfortable. We suggest that PL should come from multiple sources and via multiple strategies as critical to addressing varied educator qualifications and experience. Further, a diversity of culturally inclusive early educational approaches is required recognising, for example, worldviews and knowledges of First Nations Peoples, and the range of ethnic and religious backgrounds that make up multicultural communities worldwide.

In addition to the more traditional ways of offering PL such as workshops and conferences, there is increasing scope to develop more contemporary, flexible approaches. These might include:

A. *Online professional learning*: (a) App training such as *Sustainability from the Start,* a free, 8-module course from the *ECE Academy App* (AppStore/Google Play), available in English, French, Swedish, and soon in Croatian, Czech and Spanish, catering to diverse global audiences and curricula (co-funded by the European Union); (b) online training such as webinars (e.g. UNESCO's ESD open courses and master classes https://www.unesco.org/en/unesco-classroom); and (c) online curriculum resources (e.g. New Zealand Ministry of Education https://tewhariki.tki.org.nz/en/teaching-strategies-and-resources/contribution/education-for-a-climate-changing-future/)
B. *Designated sustainability mentors, leaders, or champions* with time release to research sustainability initiatives and support staff is becoming evident (see Davis & Elliott, 2024). These are forms of in-house PL that offer flexibility to directly target staff, service and community sustainability interests or concerns. Possibilities could involve individual mentoring, group staff discussions, related visits, or practical experiences. Such emerging strategies are invaluable in overcoming policy-practice gaps and speeding sustainability uptake in curricula.
C. *Homegrown professional networks* in Australia across six states actively share resources and expertise, in a model of distributed learning that has been successful for at least two decades. We suggest that scaling up of localised professional networks is essential across geographically and culturally diverse international early childhood communities of practitioners, researchers, and allied stakeholders. Such collaborative exchanges build new skills, knowledges, and understandings, promoting educator confidence in addressing ECEfS with young children and their communities.

6.3.4 Challenge 4: Expanding and Utilising Research as Provocations for Critical Reflection and Change

We consider research to be vital in having a skilled and knowledgeable EC workforce, and that all education and training must be informed by a robust ECEfS research agenda (Chap. 4). However, from our extensive experience in leading and analysing research and research trends (see Davis & Elliott, 2014; Elliott et al., 2020), we know there remain significant research gaps: First Nations connections and ECEfS practice and theorising; research from the 'missing' continents—Africa, South America, Asia, and the Middle East; and inclusion of children and families with diverse needs. There is also scope for further theoretical and methodological innovation (see Chap. 4). Three aspects we prioritise with examples are listed.

A. Recognising practitioners as researchers and co-researchers integral to shared critical reflection and changes in thinking and practice. This may include reflection around local meanings of sustainability and how ECEfS curricula might be implemented in specific contexts.
B. Introducing newer theoretical perspectives through research exploring post-humanism (Weldemariam, 2020) and post-materialism (Rooney & Blaise, 2023) to challenge practitioners' worldviews and values (see Chaps. 3 and 4).
C. Children's voices are integral to ECE and ECEfS, yet their voices in ECEfS research are somewhat limited (Christensen et al., 2024), particularly those aged birth to under three years. There are opportunities to promote child-centred research with a diversity of innovative methods including mosaic approach, walking interviews, child-led projects, and digital mapping studies.

6.3.5 Challenge 5: Engaging EC Corporations and Associations

In Australia, as in other countries, the ECE field is mostly run by large private for-profit corporates and small to medium businesses. These are both social enterprises with a brief to educate young children, and business enterprises aimed at making a profit for shareholders. It is imperative for businesses to show leadership in their green credentials as there are financial, reputational, and corporate risks around sustainability matters—i.e. *How many centres are in flood-prone areas or risky fire zones? What are the resource procurement practices of these businesses? Are products and resources used in centres and services eco-friendly and child-labour free? Are corporation management bodies engaging with best/next practice for sustainability?*

As Antonio Guterres (UN secretary General) comments, 'We urgently need every business, investor, city, state, and region to walk the talk on their net zero promises. We cannot afford slow movers, fake movers, or any form of greenwashing'. While it is relatively easy to create 'green' annual reports about all that a corporation is doing for the environment/sustainability, what are these actions as a proportion of their whole operation? It must not be left to individual centre managers and educational staff and children, to carry the burden of sustainability initiatives.

We note that virtually all EC management bodies offer professional learning for members. If all had viable sustainability policies, we would see a dramatic shift nationally and internationally. Elliott et al. (2016) noted in an Australian study that very few had any sustainability policy ideas. As is often the case, management policy lags contemporary sustainability initiatives in the field. Here we showcase two examples that are leading in the development of 'green' strategies and policies.

A. The European Early Childhood Education Association (EECERA) in September 2023 launched its *EECERA Sustainability SIG Position Statement* through its Special Interest Group (SIG). This marks a significant leap forward for this long-running research organisation.
B. The London Early Years Foundation (LEYF) (a community organisation of over 120 years) has developed its social enterprise and sustainability strategy, supported by a sustainability manager. The strategy's purpose is to weave its sustainability approach throughout the entire organisation, while working in partnership with the EC sector to lead and amplify best practices (see https://www.leyf.org.uk/).

6.4 New Horizons: Ideas for the Future

In this final section, we summarise our shared key ideas from our collective 100 years of teaching, researching, and advocating for ECEfS. We conclude both the chapter and the book by suggesting where/what next for ECEfS as we continue to strengthen the field through further amplification and integration of achievements.

- **Link up with wider sustainability social movements**: Support children's and young peoples' rights for activism to guide them on paths to being 'planetary' change agents (Brinkmann, 2022, p. 13). For example, support Greta Thunberg's (School Strike for Climate) movement; and engage with parents and community groups with an interest in climate change and sustainability, e.g. in Australia, Parents for Climate Change, Knitting Nanas.
- **Strengthen interdisciplinary/transdisciplinary/intersectoral links:** Reach out beyond those professionals and groups usually connected to ECE. Consider, for example, workers and organisations with climate change and sustainability links such as Doctors for the Environment, disaster resilience workers, sustainability and children's rights advocates; First Nations Peoples; and those proposing non-Western perspectives on sustainability and nature connections.
- **Create and strengthen networks:** Join existing networks such as those in Australia or internationally to support capacity building for ECEfS, locally, nationally, regionally, and globally. Grow your own ECEfS network or group offering bottom-up activism in the absence of institutional or government guidance.
- **Normalise EC educators as advocates and activists**. Demand change in ECE through writing submissions to policy reviews such as occurred in Australia in advance of the revised national curriculum; and lobby universities and vocational education providers for coherent sustainability programs in teacher education and training.
- **Build on the global nature play movement** to create synergies for sustainability. This may require critical and ethical relational changes informed by theory and research around nature play pedagogy (see Hughes et al., 2022).

- **Activate digital learning and social media** as enablers for children's and educators' learning in, about, and for sustainability. This offers much potential for broader community advocacy (see Bird & Pearson, 2024).
- **Create and continue to build ECEfS resource collections**. International online examples include those initiated by OMEP and UNESCO. Encourage national professional associations to expand their ECEfS offerings such as Early Childhood Australia's *Box of Provocations for Our Earth* https://www.earlychildhoodaustralia.org.au/our-publications/eca-special-titles/. Additionally, the USA's Children and Nature Network offers resources for reconnecting children with the natural world (https://www.childrenandnature.org).
- **Prioritise Sustainability Action Plans (SAP), policies, and position statements**. If all services, centres, associations, and corporations were to develop such guiding documents, we believe that systemic commitment to sustainability would lead to sweeping, grassroots change in ECEfS.

Finally, for us, it's time to step back, inviting next generations to take the field forward. Perhaps the next shift for ECEfS is not being seen as a 'field' at all with boundaries and permission to enter—but as a 'force field', a worldwide movement of people no matter what their background who share similar beliefs, ideas, and aims about children, early education, and living sustainability now and into the future. While we will, no doubt, continue to be involved—after all, it's been such a major part of each of our identities for so long!—there are others already present or emerging who have new things to say, in new ways, who have fresh energy and newer mindsets to speak to younger, more diverse educators and researchers in ECEfS across the globe. We wish you well!

References

Akyol, T., Pamuk, D., & Elmas, R. (2018). Drama in education for sustainable development: Preservice preschool teachers on stage. *Journal of Education and Learning, 7*(5), 102–115.

Alici, Ş. (2018). *Investigating the impact of professional development on Turkish early childhood teachers' professional growth about education for sustainable development through critical media literacy* [Doctoral thesis, Middle East Technical University, Ankara, in Türkiye].

Alici, Ş. (2020). The place of ECEFS in the Turkish teacher education system in in Türkiye. In S. Elliott, E. Ärlemalm-Hagsér & J. Davis (Eds.), *Researching early childhood education for sustainability: Challenging assumptions and orthodoxies* (Vol. 2, pp. 192–204). Routledge.

Almeida, S., et al. (2024). Sustainability education in early childhood education contexts: Insights and alternative perspectives from India, Japan, China and Singapore. In J. Davis & S. Elliott (Eds.), *Young children and the environment: Early learning for sustainability* (3rd ed.). Cambridge University Press.

Ärlemalm-Hagsér, E., Engdahl, I., & Pramling Samuelsson, I. (2023). Förskollärares mångfaseterade motiv för undervisning om hållbarhet [Preschool teachers multiple motives for teaching about sustainability]. *Nordisk barnehageforskning, 16*(3), 1–22. https://doi.org/10.23865/nbf.v19.345

Ärlemalm-Hagsér, E., Gustavsson, L., Thulin, S., & Evans, N. (2022). New Initiatives for Building Education for Sustainability in Initial Early Childhood Teacher Education in Sweden – Critical Aspects and Noticeable Needs. *Australian Journal of Teacher Education, 47*(5), 67–79.

Arts, K. (2019). Children's rights and climate change. In C. Fenton-Glynn (Ed.), *Children's rights and sustainable development: Interpreting the UNCRC for future generations*. Cambridge University Press.

Australian Government Department of Education (AGDE). (2022). *Belonging, being and becoming: Early years learning framework for Australia* (V2.0). Australian Government Department of Education for the Ministerial Council.

Baker, C., Clayton, S., & Bragg, E. (2021). Educating for resilience: Parent and teacher perceptions of children's emotional needs in response to climate change. *Environmental Education Research, 27*(5), 687–705. https://doi.org/10.1080/13504622.2020.1828288

Bansal, A., et al. (2023). Heatwaves and wildfires suffocate our healthy start to life: Time to assess impact and take action. *Lancet Planet Health, 2023*(7), e718–e725.

Berry, A. M. (2022). Weaving child-plastic relations with early childhood educators in the Ecuadorian Andes. *Australian Journal of Teacher Education, 47*(5). https://doi.org/10.14221/ajte.2022v47n5.6

Bird, J., & Pearson, M. (2024). Harnessing children's interest in digital technologies for sustainability learning: Catalyst for conversation. In J. Davis & S. Elliott (Eds.), *Young children and the environment* (3rd ed.) (pp. 189–209). Cambridge University Press.

Brinkmann, R. (2022). Defining sustainability. In C. Constance (Ed.), *The Palgrave handbook of global sustainability*. Springer Nature.

Christensen, M., O'Gorman, L., Walsh, K., & Davis, J. (2024). Missing voices: Children's understandings of sustainability topics and the 4C Pedagogical Framework. In J. Davis & S. Elliott (Eds.), *Young children and the environment: Early learning for sustainability*. Cambridge University Press.

Clark, H. et al., (2020). A future for the world's children? A WHO–UNICEF–Lancet Commission. *The Lancet Commissions, 395*(10224), 605–658. https://doi.org/10.1016/S0140-6736(19)32540-1

Clarke, A. (2023). *Slow knowledge and the unhurried child: Time for slow pedagogies in early childhood education*. Routledge.

Cooke, S., & Baldwin, N. (2024). Healthy and sustainable environments for children and communities in a changing climate. In J. Davis & S. Elliott (Eds.), *Young children and the environment: Early learning for sustainability*. Cambridge University Press.

Cutter-Mackenzie, A., Edwards, S., Moore, D., & Boyd, W. (2014). *Young children's play and environmental education in early childhood education*. Springer.

Davis, J. (2014). Examining early childhood education through the lens of education for sustainability: Revisioning rights. In J. Davis & S. Elliott (Eds.), *Research in early childhood education for sustainability: International perspectives and provocations* (pp. 21–37). Routledge.

Davis, J. M., & Davis, E. D. (2020). Early childhood teacher education and education for sustainability: A review of the literature and mapping of courses. In S. Elliott, E. Ärlemalm-Hagsér, & J. Davis (Eds.), *International perspectives on early childhood education for sustainability: Challenging assumptions and orthodoxies* (pp. 111–124). Routledge.

Davis, J., & Elliott, S. (2024). *Young children and the environment* (3rd ed.). Cambridge University Press.

Davis, J., & Elliott, S. (Eds.). (2014). *Research in early childhood education for sustainability: International perspectives and provocations*. Routledge.

Dreamson, N., & Kim, S. (2022). Human–nature relationships in early childhood education for sustainability: Metaphysical analysis and reshaping. *Australasian Journal of Early Childhood, 47*(2), 121–134.

Elliott, S., & Davis, J. (2023). *Early childhood Australia essentials: Sustainability in early childhood. Almost mainstream: Re-imagining early childhood education for sustainable futures*. Early Childhood Australia.

Elliott, S., & Pugh, R. (2020). Children's voices about fish and tadpoles in an Australian pond ecosystem: It's all about balancing and belonging. In S. Elliott, E. Ärlemalm-Hagsér, & J. Davis (Eds.), *Researching early childhood education for sustainability: Challenging assumptions and orthodoxies* (Vol. 2, pp. 205–219). Routledge.

Elliott, S., & Davis, J. (2009). Exploring the resistance: An Australian perspective on educating for sustainability in early childhood. *International Journal of Early Childhood, 41*(2), 65–77.

Elliott, S., Ärlemalm-Hagsér, E., & Davis, J. (Eds.) (2020). *Researching early childhood education for sustainability: Challenging assumptions and orthodoxies* (Vol. 2). Routledge.

Elliott, S., McCrea, N., Newsome, L., & Gaul, J. (2016). *Examining environmental education in NSW early childhood education services: A literature review with findings from the field. NSW OEH Environmental Trust.* http://www.environment.nsw.gov.au/grants/dissemination.htm

Evans, N., Harwood, D., Furu, A.-C., & Alici, Ş. (2022). Challenges in reorienting early childhood education for sustainability in initial teacher education: Transforming, creating and expanding. *The Australian Journal of Teacher Education, 47*(5). https://doi.org/10.14221/ajte.2022v47n5.1

Farrant, B., Armstrong, F., & Albrecht, G. (2012). *Future under threat: Climate change and children's health.* The Conversation. https://theconversation.com/future-under-threat-climate-change-and-childrens-health-9750

Fenton-Glynn, C. (Ed.). (2019). *Children's rights and sustainable development: Interpreting the UNCRC for future generations.* Cambridge University Press.

Ferreira, J., Ryan, L., & Davis, J. (2021). A deep dive into sustainability leadership through education for sustainability. In F. English (Ed.), *The Palgrave handbook of educational leadership and management discourse* (pp. 1–21). Palgrave Macmillan.

Helldén, D., Andersson, C., Nilsson, M., Ebi, K. L., & Fribe, P. (2021). Climate change and child health: A scoping review and an expanded conceptual framework. *Lancet Planetary Health, 5*(3), 164–175.

Hughes, F., Elliott, S., Anderson. K., & Chancellor, B. (2022). *Early years learning in Australian natural environments.* Oxford University Press.

Hughes, F. (2021). *Problematising sustainability in nature play programs: Pedagogical transformations and ambitious alignments* [Doctor of Philosophy, University of New England].

Ifous. (2021). *Plan för forskning-och utvecklingsprogrammet: Hållbar förskola.* [Plan for the research and development programme: Sustainable preschool]. Ifous.

Inoue, M. (2014). Perspectives on early childhood environmental education in Japan: Rethinking for a sustainable society. In J. Davis & S. Elliott (Eds.), *Research in early childhood education for sustainability: International perspectives and provocations* (pp. 79–96). Routledge.

Jannok-Nutti, Y. (2010). Ripsteg mot spetskunskap i samisk matematik: lärares perspektiv på transformeringsaktiviteter i samisk förskola och sameskola [*Grouse steps towards front line knowledge in Sámi mathematics: Teachers' perspective on transformations activities in Sámi preschool and Sámi school*] [Thesis, Luleå technical University].

Ji, O., & Stuhmcke, S. (2014). The project approach for early childhood education for sustainability: Exemplars from Korea and Australia. In J. Davis & S. Elliott (Eds.), *Research in early childhood education for sustainability: International perspectives and provocations* (pp. 158–179). Routledge.

Kagawa, F. (2022). *Towards climate resilient education systems. A tool for reflection, dialogue and progress assessment for Ministries of Education and its partners.* UNICEF Regional Office for South Asia. https://www.unicef.org/rosa/media/18011/file/Towards%20Climate%20Resilient%20Education%20Systems.pdf

Kenis, A., & Mathijs, E. (2012). Beyond individual behaviour change: The role of power, knowledge and strategy in tackling climate change. *Environmental Education Research, 18*(1), 45–65.

Laiti, M., & Määttä, K. (2022). An ecocultural consideration of Sámi early childhood education. *European Journal of Education Studies, 9*(2), 62–78. https://https://doi.org/10.46827/ejes.v9i2.4149

Laiti, M., Määttä, K., & Köngäs, M. (2022). Sámi early childhood education and sustainability in the Arctic. *International Journal of Research in Education and Science, 8*(4), 783–799. https://doi.org/10.46328/ijres.2974

Mackey G., & Ritchie J. (2012). *Pedagogies for Sustainabilities in Early Childhood Education in Aotearoa New Zealand.* York University, Toronto, Canada: Seventh Biennial Meeting of the International Network of Teacher Education Institutions UNESCO, May 16–20, 2012.

Mezirow, J. (1997).Transformative learning theory. *New Directions for Adult and Continuing Education, 74*, 5–12.

NZ Ministry of Education. (2017). *Te Whariki.* NZ Ministry of Education.

Plumwood, V. (2002). *Environmental culture: The ecological crisis of reason.* Taylor & Francis.

Pramling Samuelsson, I., & Kaga, Y. (Eds.). (2008). *Early childhood and its contribution to a sustainable society.* UNESCO.

Ritchie, J. (2014). Learning from the wisdom of elders. In J. Davis & S. Elliott (Eds.), *Research in early childhood education for sustainability: International perspectives and provocations* (pp. 36–49). Routledge.

Rooney, T., & Blaise, M. (2023). *Rethinking environmental education in a climate change era: Weather learning in early childhood.* Routledge

Sanson, A. V., Van Hoorn, J., & Burke, S. E. L. (2019). Responding to the impacts of the climate crisis on children and youth. *Child Development Perspectives, 13*(4), 201–207. https://doi.org/10.1111/cdep.12342

Spiteri, J. (2021). Can you hear me? Young children's understanding of environmental issues. *International Studies in Sociology of Education, 30*(1–2), 191–213. https://doi.org/10.1080/09620214.2020.1859401

Swedish National Agency for Education. (2019). *Curriculum for the preschool Lpfö18.* Translated to English. https://www.skolverket.se/publikationer?id=4049

Taylor, A. (2013). *Reconfiguring the natures of childhood.* Routledge. https://doi.org/10.4324/9780203582046

Taylor, C. A. (2016). Edu-crafting a cacophonous ecology: Posthumanist research practices for education. In C. A. Taylor & C. Hughes (Eds.), *Posthuman research practices in education* (pp. 5–24). Palgrave Macmillan.

Taylor, A. (2017). Beyond stewardship: Common world pedagogies for the Anthropocene. *Environmental Education Research, 23*(10), 1448–1461. https://doi.org/10.1080/13504622.2017.1325452

Tilbury, D., & Mulà, I. (2023). *Learning from thirty years of experience: Case studies in teacher education for sustainability, EENEE report.* Publications Office of the European Union.

IPCC UNEP. (2023). *Climate change 2023: Synthesis report.* UNEP-UN Environment Programme.

United Nations (UN). (2015). *Transforming our world: The 2030 agenda for sustainable development.* https://sdgs.un.org/publications/transforming-our-world-2030-agenda-sustainable-development-17981

United Nations Committee on the Rights of the Child (CRC). (2023). General comment No. 26 on Children's Rights and the Environment. https://www.ohchr.org/en/documents/general-comments-and-recommendations/crccgc26-general-comment-no-26-2023-childrens-rights

United Nations Education, Scientific and Cultural Organization. (2017). *Education for sustainable development goals: Learning objectives.* UNESCO. https://www.unesco.de/sites/default/files/2018-08/unesco_education_for_sustainable_development_goals.pdf

Weldemariam, K. (2020). Reconfiguring environmental sustainability in early childhood education: A post-anthropocentric approach. *Environmental Education Research, 26*(12), 1789–1789. https://doi.org/10.1080/13504622.2020.1854690

Weldemariam, K. (2022). Broadening ontological and epistemological possibilities within early childhood teacher education for sustainability. *The Australian Journal of Teacher Education, 47*(5), 33–49. https://doi.org/10.14221/ajte.2022v47n5.3

Weldemariam, K., Boyd, D., Hirst, N., Sageidet, B. M., Browder, J. K., Grogan, L., & Hughes, F. (2017). A critical analysis of concepts associated with sustainability in early childhood

curriculum frameworks across five national contexts. *International Journal of Early Childhood, 49*(3), 333–351.
Wong, M. (2019). Inclusion of Indigenous children's rights. In C. Fenton-Glynn (Ed.), *Children's rights and sustainable development: Interpreting the UNCRC for future generations*. Cambridge University Press.

Key Terms

Anthropocene: Newly developed term identifying the period from the industrial revolution to the present when human activity on Earth is deemed to have become so significant as to have dramatically reshaped Earth and its biological and ecological systems.

Early childhood education (ECE): Represents care and education processes for that period in the human life span from Birth to 8 years.

Early childhood education for sustainability (ECEfS): The synergy of ECE and education for sustainability into a unique field of educational research, policy, and practice.

Ecocene: Proposed time period identifying the age following the Anthropocene during which human activity will restore and regenerate, rather than continue to negatively impact the Earth.

Environmental education (EE): Holistic, lifelong learning approach aimed at creating eco-responsive citizens with abilities to explore, identify, and take action to solve environmental issues. Originally, it focused on education in, about, and for the environment, now mostly subsumed into EfS.

Education for sustainability (EfS)/Education for sustainable development (ESD): Develops knowledge, skills, ethics, and worldviews for people to act in ways that contribute to sustainable ways of living now and into the future. ESD is mostly referenced by the United Nations and European countries. In Australia, EfS is the more common terminology.

First Nations Peoples (FNP): Originating in Canada, this term has become more widely used across the globe to identify the original inhabitants of colonised lands. Alternative terms include Indigenous peoples, or Aboriginal and Torres Straits Islander peoples (Australia), though all such terms are not without controversy and debate.

Initial teacher education (ITE): A contemporary term to describe what has been referred to as pre-service teacher education/teacher training. Typically, it confers a university degree qualification to prepare students for a career in teaching.

J. M. Davis et al., *Early Childhood Education for Sustainability*, SpringerBriefs in Education, https://doi.org/10.1007/978-981-97-8405-9

Intergenerational equity: Offers an extension to discussions about equity that calls for equitable distribution of Earth's resources between current and future generations so that future generations are no worse off than people living today.

Nature education/nature play: These terms encompass learning in, about, and from natural environments. They have a strong tradition in many early childhood education and care settings.

Post-humanism: Thinking that challenges the notion of humans as exceptional and privileged in relation to non-human beings, such that human needs and wants override and often destroy the non-human. Humans, instead, are seen as physically and biologically enmeshed and dependent on non-human species and environments.

Sustainability/Sustainable development (SD): Contentious because of their anthropocentric orientations, these concepts refer to societal development aimed at simultaneously meeting the needs of current generations without compromising those of future generations. Sustainability is more commonly used in Australia, while in the USA and Europe, SD is more common. Sustainability is often described as having between three to five interrelated dimensions: social-cultural-political, economic, and natural.

Sustainable Development Goals (SDGs): The 17 goals identified by the United Nations provide guidance for addressing key global challenges including poverty, inequality, climate change, biodiversity, peace, and justice.

Systems thinking/systems change: These ideas emerged as responses to reductionist problem-solving that has failed to sufficiently cope with the realities of complexity and uncertainty in social and ecological systems. Systems approaches identify connections and synergies offering novel, more realistic ways of framing and overcoming problems and issues, thus helping to break through inertia, barriers, and resistances to change.

Transformative/transformational learning: This type of learning involves processes of deep, constructive, and meaningful learning that goes well beyond knowledge acquisition. It supports critical thinking and offers learning to create fundamental worldview changes.

Whole school approach (WSA): This approach to education is an example of systems thinking where the whole of an educational setting or institution is engaged in developing and implementing cohesive, collaborative, and transformative approaches across all elements of the system—curriculum, pedagogies, leadership, management, human and physical environments, and community connections.

The manufacturer's authorised representative in the EU is Springer Nature Customer Service Centre GmbH, Europaplatz 3, 69115 Heidelberg, Germany. If you have any concerns regarding our products, please contact ProductSafety@springernature.com

Printed and bound by CPI Group (UK) Ltd, Croydon, CR0 4YY
01/10/2026
02225839-0009